ELECTRONIC PUBLISHING FOR PHYSICS AND ASTRONOMY

ASTROPHYSICS AND
SPACE SCIENCE LIBRARY

VOLUME 224

ELECTRONIC PUBLISHING FOR PHYSICS AND ASTRONOMY

Edited by

ANDRÉ HECK

Strasbourg Astronomical Observatory, France

Reprinted from *Astrophysics and Space Science*
Volume 247, Nos. 1–2, 1997

KLUWER ACADEMIC PUBLISHERS

DORDRECHT / BOSTON / LONDON

A C.I.P. Catalogue record for this book is available from the Library of Congress.

ISBN-13: 978-94-010-6514-6   e-ISBN-13: 978-94-009-0055-4
DOI: 10.1007/978-94-009-0055-4

Published by Kluwer Academic Publishers,
P.O. Box 17, 3300 AA Dordrecht, The Netherlands.

Sold and distributed in the U.S.A. and Canada
by Kluwer Academic Publishers,
101 Philip Drive, Norwell, MA 02061, U.S.A.

In all other countries, sold and distributed
by Kluwer Academic Publishers,
P.O. Box 322, 3300 AH Dordrecht, The Netherlands.

Printed on acid-free paper

TABLE OF CONTENTS

FOREWORD

The ship has left the Miraflores Locks, let loose from the 'mules' run by the crews of the Panama Canal Commission. She has picked up speed while passing under the Bridge of the Americas which links *de facto* the Northern and Southern parts of the continent, and has headed resolutely towards the Pacific Ocean waters along the rows of boats of all kinds waiting to cross the Canal in the other direction. Through a layer of tropical clouds, the setting Sun is bleakly illuminating the tall white highrises of Panama City on the port side. It took a full day to cautiously move through the whole system of locks and cuts.

Back in the stateroom, I open again a working copy of this book and type down this foreword on the pocket computer. The last chapter was received the day before while speeding through the Caribbean Sea and my main work as Editor is now over.

It has been a real pleasure and a great honour to be given the opportunity of compiling this book and interacting with the various contributors through the latest technologies while being sometimes in geographically very different places.

The quality of the authors, the scope of experiences they cover, the messages they convey make of this book a unique and timely publication. The reader will certainly enjoy as much as I did going through such a variety of well-inspired chapters from so many different horizons.

It is not often the case either that publishers detail their activities and projects in a publication by one of their competitors. In this respect, it is also a very pleasant duty to pay tribute here to the various people at *Kluwer Academic Publishers* who quickly understood the interest of such a volume and enthusiastically agreed to produce it.

The volume starts with a few general chapters. After a contextual introduction by the Editor, Arnoud de Kemp offers his deep experience with scientific, technical and medical publishing while Franco Mastroddi details the challenge of electronic publishing (EP) for the multilingual and multi-

Astrophysics and Space Science **247**: vii–viii, 1997.
© 1997 *Kluwer Academic Publishers.*

cultural European mosaic. Fionn Murtagh then describes the technologies called for by EP.

Then follow a series of contributions from officers of learned societies or from persons dealing with their publications. Benjamin Bederson and Harry Lustig from the *American Physical Society* explain the rôle of a large scientific society as far as EP is concerned, while Peter Boyce from the *American Astronomical Society* recalls the activities of a smaller – but no less pioneering – professional group. Harm Habing and James Lequeux describe the multinational European venture in EP for *Astronomy & Astrophysics*. As a reference, the very detailed EP plan and interim copyright policy of the *Association for Computing Machinery Inc. (ACM)* are reproduced here under the original joint authorship of Peter Denning and Bernard Rous. This section is concluded by the chapter of Dennis Shaw on the EP programme at the *International Council of Scientific Unions (ICSU) Press*.

We then move to the delicate matters of economic issues with Harry Lustig, as well as of copyright and protection for electronic material with Douglas Armati and Edward Barrow.

Uta Grothkopf quite naturally follows with her views from her key position of librarian having to cope with all the changes in the scientific information world.

Then come a group of chapters from cosmic information providers as a foretaste of what will be the information hubs of the future: Daniel Egret and Françoise Genova share their experience from the *Strasbourg Astronomical Data Centre (CDS)* while Guenther Eichhorn details the development of the *NASA Astrophysics Data System (ADS)*. The *Star*s Family* resources are then described by André Heck as an EP example of validated and authenticated yellow-page services.

Last, but far from being least, two chapters from publishing groups conclude the volume: Anne Dixon details the development of EP at the *Institute of Physics Publishing* while Johannes Menzel, Ken Metzner and Elizabeth Pope present two ambitious projects at *Academic Press*.

Certainly there were several possible sequences for presenting the various contributions of this compendium. None would be perfect since each specific chapter belongs generally to several categories. There is no simple truth nor single answer either to the various issues and challenges raised by EP, so do not expect all the authors to be 'on the same wavelength'. One thing is sure however: all these chapters and the underlying experiences are worth perusing and meditating. Take advantage of these treasures!

André Heck
'Legend of the Seas'
7 April 1997.

ELECTRONIC PUBLISHING IN ITS CONTEXT

A. HECK
Observatoire Astronomique
11 rue de l'Université
F-67000 Strasbourg, France
heck@astro.u-strasbg.fr
http://cdsweb.u-strasbg.fr/~heck

Abstract. Contextual aspects of electronic publishing (and more generally of diversified publishing) are discussed. Definitions and concepts are introduced. Pending issues and challenges are identified. The accent is put on the need for providing authenticated and validated information. The electronic medium is a new medium *per se* that will exist together with other ones, such as paper, but it will call for specific procedures, strategies and policies. A lot has still to be done on the human level.

1. Introduction

Electronic publishing (EP) is slowly taking shape. The technology is there, certainly evolving rapidly and will progress still further in the future, but a number of factors have dilatory effects and are sometimes underestimated. If it is obvious that advantage has to be taken of the technological advances, it is not yet quite clear for many of us what will be the exact impact of these, nor in fact whether all the potentialities at hand are well understood. Human inertia is also a very limiting agent and, speaking of scientific communities, the traditional procedures and the habits progressively adopted over the decades (if not centuries) are difficult to alter. Human nature is simply and basically reluctant to change, especially when the final outcome is not well or not fully perceived.

The physics and astronomy communities have been among the first involved in EP, even before the concept itself existed *per se*. Astronomers, space physicists, high-energy physicists and their colleagues around the world have done more than just help in setting up the Internet and the

Astrophysics and Space Science **247**: 1–10, 1997.

associated networks. They jumped onto the World-Wide Web (WWW) and quickly became prolific producers and eager consumers of its resources (Hardin 1993).

As scientists, our ultimate aim is to contribute to a better understanding of, *stricto sensu*, the universe (as well as of its past and future) and consequently to a better comprehension of the place and rôle of man in it. To this end, together with theoretical studies, we carry out observations to obtain data that will undergo treatments and studies leading to the publication of results. The whole procedure can include several iterations or interactions between the various steps as well as with external fields, non-scientific disciplines, instrumental technologies, and information handling methodologies

Electronic information handling is a broad and flexible concept with plenty of degrees of freedom, adapted to the fluid and living nature of today's information material. It encompasses data collection, analysis, dissemination, and so on, as well as publishing (classical or otherwise). All these elements cannot be dissociated from each other as the electronization has facilitated the various interactions upstream and downstream, the ultimate step being a *diversified publication* of the final results. The classical scheme involving authors, editors, referees, publishers and readers – or, more generally speaking, information providers and users plus intermediaries – is also changing and can become very complex if including authentication and validation loops. It should be kept in mind that publishing is not only motivated by information sharing, but also strongly conditioned by career constraints.

There have not been many dedicated conferences on electronic publishing. In the general field of science, ICSU Press and UNESCO have to be credited with an impressive meeting (Shaw and Moore 1996; Shaw 1997). There have been also a number of events in the rather compact and well-structured astronomical community (see e.g. Heck 1992; Heck & Murtagh 1996)[1].

2. Definitions and concepts

In the following, *information* will be considered as what is communicated by others or obtained from investigation, study, or instruction. It covers the observational material, the more or less reduced data extracted from it, the scientific results, as well as the accessory material increasingly used

---

[1] Astronomy has also had regular conferences and reference publications on electronic information handling. Refer for instance to the various contributions in Egret & Albrecht (1995), Egret & Heck (1995), Heck & Murtagh (1993), Murtagh *et al.* (1995) and to the references quoted therein.

by scientists in their work (bibliographical resources, yellow-page services, software libraries, and so on).

While *publication* will be considered as a public announcement (no implicit assumption being made as to the medium used), *communication* will be taken as the act or action of imparting or transmitting (while respecting constraints such as proprietary rights and so on). These definitions correspond to widely accepted concepts in information sciences, but the meaning of the terms could be different in fields such as marketing or advertising. Of course, it should also be kept in mind that data analysis, information sharing and related activities should never be an end *per se* as science must remain the main objective.

The structure itself of information has become different: beyond the classical quasi-linear layout of publications on paper, electronic documents include *hypertextual links*, the structure of which is more closely adjusted to the mental structure of many people.

The information material as a whole is now existing in an increasingly distributed way. Data centres have seen their rôle evolving and tend now to act more as *hubs* towards distributed specialized repositories of different types of data and material (rather than, as in the past, holding as much as possible themselves and carrying out the integration work on their very location). This is thanks largely to the fact that the evolution of information technology has brought major modifications in relation to hardware and connectivity as well as new tools (client/server facilities, WWW browsers, resource discovery packages, ...) and concepts (hypertext/hypermedia concepts, virtual libraries, ...).

Finally, we have now entered for good the era of *fluid information*, *i.e.* a material that can be continuously updated, upgraded, enlarged, improved, modified, and so on. This new concept implies those of *document (in)stability* and of *document genetics* : beyond its own permanent possible evolution, a document can give birth to subsidiary ones, first linked to itself; the relevance of some of these can then supplant with time that of the original document that would virtually 'die'. Forgetting this fluidity would be equivalent to staying with CD-ROMs, which are frozen repositories of fixed information and sometimes remain short of answering adequately some of our current needs.

3. The new medium

The emergence of the electronic medium is currently best represented by the WWW (but what will it be to-morrow? microelectrodes linked to a bio-cyberspace? – see *e.g.* Gibson 1986 & 1993). The WWW is based on hypertext and hypermedia. It has become, with unprecedented speed, a magnificent communication tool that has been called the *'fourth media'* and which is *de facto* a fantastic cross-disciplinary, cross-educational and cross-social meeting ground allowing exchanges on a new dimension. It is a highly dynamic domain, evolving rapidly.

Each of us has become an actual or potential author-creator of electronic documents acquiring *ipso facto* very rapidly an extremely high visibility, well beyond the horizon traditionally reached in specific circles. This is specially due to the tremendous efficiency of the formidable search tools available on the web. One must then be fully conscious of this and prepare in line with the consequences (*i.e.* with *ad hoc* caution and ethics) any document to be published.

The best search engines include *Yahoo* (URL: http://www.yahoo.com), *Lycos* (URL: http://www.lycos.com), and Digital's *Alta Vista* (URL: http://www.altavista.digital.com) which has our preference. An example of resources with search engines more specific to astronomy and related fields is described elsewhere in this volume (Heck 1997).

The explosion of electronic documents is however not without bringing in new questions, new challenges and new problems that will have to be faced, especially on the ethical, legal and educational levels, without forgetting the security nor the fragility of the material delivered on the electronic medium. We shall mention only a few specific points here. The interested reader will find more details in Heck (1995 & 1996) and in the references quoted therein.

4. Diversified publishing

Flexible publishing or, as we prefer calling it, *diversified publishing* implies that we shall be able to go to any media we like (WWW, CD-ROM, paper, and so on) in a hopefully automated fashion. The road to achieve this fully requires *i.a.* to make producing multimedia faster and more efficient.

But there are still too frequent timorous and/or conservative attitudes in view of what is possible with the current development of technologies and methodologies. Too many people remain short of the potentialities of the new medium and see still electronic publishing as little more than putting on line an electronic version of something that is existing also on paper.

Do not misunderstand the above though. Putting on line a printed document is not wrong, but this is by far insufficient. Why? Simply because the electronic medium is exactly what that means, a new medium *per se*, complementary to the existing ones, and because its usage should imply – and even require – dedicated techniques, policies, and strategies.

This is such an obvious statement that it probably does not need lengthy illustrations. Comparisons are often drawn with the advent of radio, or better television. The introduction of a new medium does not lead to the disappearance of former ones (in this case, newspapers and magazines). It calls however for a specific approach tailored for it in the same way that, on TV, they do not zoom in on newspapers or broadcast people reading magazines.

It is obvious however that electronic-publishing policies have not yet reached a final degree of maturity in spite of the fact that some of them are already quite elaborated (see various chapters of this volume and the references therein). Few of these policies go beyond an 'electronization' of a paper document and of the previous procedures used to deal with it. Certainly these are made faster and more flexible, but still fall short of satisfactorily providing a solution for the fluid nature of today's information and the living character of information retrieval on the WWW.

It must be clear that, wishing to maintain at all costs a compatibility between PostScript, PDF and HTML (something that is repeatedly announced in publishing ventures) would prevent taking advantage of the hypertextual structure, sound, motion, applets and whatever may come next, available on the electronic medium.

Thus the often-heard opposition between classical publishing on paper and electronic publishing is unfounded, since, as explained above, the new medium is complementary of existing ones. The latter will have to adjust themselves to the arrival of the newcomer (as newspapers and magazines had to do for television), but there is no reason for publications on paper to disappear.

5. Validation and authentication

It is also obvious that learned societies, funding organizations, expert committees and other bodies will have to integrate the diversified-publishing productions into their *evaluation* procedures, that is the assessment of activities, plans and projects of individuals and organizations, implying a revision of the corresponding practicalities for *recognition*.

Indeed the phenomenology of publishing is not only motivated by the need of sharing information, but also strongly conditioned by recognition, a necessity that should not be underestimated and which is largely based on publications in refereed journals. This is sought for getting positions (grants and salaries), for obtaining acceptance of proposals (leading to data collection), and for achieving funding of projects (allowing materialization of ideas).

This implies of course another step: the adaptation of *validation* procedures to electronic material ('refereeing' it) and of measures for guaranteeing subsequently its integrity (see below). Reliable validation procedures are more than ever necessary because it has become increasingly difficult to distinguish between the so-called grey literature and the formal one.

Additionally *authentication* of originators (authors, institutions or organizations) might become an increasingly critical issue with the electronic material, as well as its importance itself being increasingly acknowledged.

A phenomenon that has to be appreciated is that, independently from any validation procedure, servers and web documents of persons and organizations with *profile* and *reputation* will be visited regularly with preference and confidence, so disrupting the current chronology of preprint-submission-publication.

One could even wonder whether servers of preprints and proceedings of conferences, not to forget those of personal documents and productions, will not take over if the procedures of the learned societies, the commercial publishers and other traditional channels remain slow and heavy, failing to respond to the dynamism, the fluidity and the visibility available via the electronic vector.

Mentalities, habits and policies will have to adjust themselves progressively, with the usual delayed reaction time resulting from natural human and social inertia, coupled to a certain reluctance, if not sometimes a definite distrust, towards the new medium, linked to the *fragility* of the electronic material and to the alterations it could easily undergo.

6. Maintenance for quality

The *maintenance* process of information resources must be continuously improved from lessons learned with time and by using the most appropriate tools. Generally speaking, information has to be collected, verified, de-biased, homogenized, and made available not only in an efficient way, but also through operationally reliable means (it becomes useless if plugged into a confidential network or reachable through deficient routers). Redundancies have to be avoided; precision is, and details can be, extremely important.

If scientists have a natural tendency to design projects and software packages involving the most advanced techniques and tools, there is in general less enthusiasm for the painstaking and meticulous long-term maintenance which builds up the real substance of the quality resources. This has also to be carried out by knowledgeable scientists or documentalists and cannot be delegated to inexperienced clerks or temporary employees, since the necessary experience is long and slow to be acquired.

Information retrieval *per se* is raising a number of *evaluation* issues (see *e.g.* Harman 1992 and the subsequent papers of the corresponding special issue). The fashion is now shifting towards designing and experimenting with *quality control* processes. This might be a very serious matter or a big joke. None of the algorithms currently available has really convinced us of their absolute necessity and satisfactory efficiency.

Such a short time after its birth, the web needs already a good *cleaning*, so numerous are the anchors pointing towards inexisting documents and so many are useless or obsolete documents on the various servers worldwide. It is certainly up the 'webmasters' to fix this and to prune the dead wood from their respective sites.

The *stability* of sites and URLs is very important. Also URLs should not be modified unless for good reasons. Forwarding pointers should be put in place, failing what it could be impossible to keep track of the moving sites.

7. Ethics, security, and copyright

At a time when authors/creators of electronic documents are increasingly worried about the easy possible alteration of their work, proper *credit* to the material used should always be clearly indicated. Since the browsers make it so easy to download the original files, crediting the sources appropriately becomes critically important. It is also smarter and more elegant to insert a hyperlink to the original document since it will point then always to the freshest version of the file.

This brings us to *security* issues, involving monitoring the visits of the server, restricting access to some documents, preserving the confidentiality of others, and so on. Away from governmental policies (such as the *Clipper chip* project in the US that has been raising substantial controversy or the current ban on encryption in France), there is no golden rule on security issues.

It is up to each local 'webmaster' to set up appropriate security. Some resources require *ad hoc* clearance (password, account number and so on); others will be only partially retrievable in a specific query (such as large copyrighted databases); finally, other documents are freely accessible and usable, conditioned on a minimum of ethical behavior (see above). Electronic commercial transactions implying transfer of funds (including using a credit card number) have initiated a number of elaborated procedures (at the limit of paranoia) trying to prevent in advance all potential tricks from computer gangsters.

It is also not excluded that an astronomical *intranet* will have to be set up at some stage in order to make more efficient the exclusively professional exchanges.

Legal aspects (copyright, electronic signatures, ...) are also extremely important and jurists are busy setting up references for the computerized material. Particularly in this case, there might be variations from country to country when the law already exists. However, with the world globalization of electronic communications, one can expect – and hope for – quick harmonization of the various references and procedures. On such matters, refer *e.g.* to specific chapters in this volume as well as to Samuelson (1994 & 1996) and, more generally, to her very interesting regular column *'Legally Speaking'* in the *ACM Communications*.

Some jurists are still discussing whether offering material on-line, displaying material on screen, and storing material temporarily are acts subjected to the usual copyright protection.

As a rule however and from a recent experience, authors should be very careful with what they sign in terms of copyright transfer agreement and possibly refuse to transfer any copyright at all. If they are not careful enough, they might hand over much more than they would normally think, up to a total exclusivity in favor of the publisher which would prevent even a posting on a personal WWW site. The only place where the contribution could be found would be the publisher's server, quite likely against payment. This would be another example of the fact that 'copyright' does not mean a protection for the author, but rather for the publisher, restricting very seriously the authors' right on – and personal promotion of – their own work.

Minimum rights for the authors have to be protected. It is thus important to request that all restrictions (if any) that would apply in terms of electronic distribution and postings of papers be precisely specified in the copyright transfer agreement itself, as well as in the instructions for authors. Guest editors should also see this mentioned in their contracts with the publishers. These conditions should not change or evolve during the publishing process, unless by mutual agreement.

One might have differing opinions on the rôle of publishers[2] and on the financial implications of the publishing process. From all the experience built up through our activities in the field of electronic publishing and information handling, our own stand is that, at the very least, an author should always be allowed to post his full papers on his own WWW site. Collaboration with publishers preventing this should be seriously questioned.

With the advent of the electronic era, scientists and scientific institutions have now all the possibilities to run, if necessary, an efficient information server with validated (refereed) material without the help of a commercial publisher. Ginsparg (1996) actually concurs on this: "A correctly configured fully electronic scholarly journal can be operated at a fraction of the cost of a conventional print journal, and could for example be fully

---

[2]We have had very differing experiences with the ones we dealt with.

supported by author subsidy (page charges or related mechanism, as already paid to some journals), ideally allowing for free network distribution and maximal benefit both to authors and readers."

The technical expertise is there, as well as a network of referees already functioning on a volunteer basis (and increasingly via electronic transfers), in such a way that only limited funding will be necessary for the clerical work of the editors and a minumum of equipment. Of course, a number of precautions will be necessary such as mirror sites, guarantees for the permanency of service and the integrity of the refereed/validated material, links to possible updates and complementary information, etc. But there is certainly no major obstacle to providing an excellent electronic-publishing resource in the full sense of the expression.

8. And the human component?

Last, but not least, there are non-negligible *educational aspects* to be taken into account as to the introduction and training of young and not-so-young people to the new technologies within the various communities. This is true not only for scientists, but also for librarians and documentalists who will see their rôle significantly changing within their institution and who will be increasingly dealing with virtual material (see *e.g.* Brown 1996 and Grothkopf 1997).

We have already mentioned how it will be necessary to adapt our habits and policies to the new electronic medium, not only in everyday life, but also for proper recognition of the scientific activities, the results of which will quite naturally be expressed on the various media (and not only on paper). Educating all human components involved in the new concepts, capabilities, methodologies and technologies is a process that will also require patience, dedication and ... time, as technology leaders agree at least on one point: change is this only sure thing in the next decade.

As reminded recently by Gell-Mann (1997), with the digital age producing an 'immense sea of data that threatens to drown humanity', people need to adapt how they think so that true knowledge can be distilled from the deluge. "We hear, in this dawn of the so-called information age, a great deal of talk about the explosion of information and new methods for its dissemination. It is important to realize, however, that most of what is disseminated is misinformation, badly organized information or irrelevant information. How can we establish a reward system such that many competing but skillful processors of information, acting as intermediaries, will arise to interpret for us this mass of unorganized, partially false material?"

We have to rely indeed on the wisdom of providers and users of electronic information, as well as on the various intermediaries (compilers, information hubs, and so on) and on the learned societies, committee experts, and so on, to take this into account, to include quickly validated and authenticated electronic information in the evaluation processes, and to give to electronic publishing its deserved *lettres de noblesse*.

References

1. Brown, D.J. (Ed.) 1996, Electronic publishing and libraries. Planning for the impact and growth to 2003, Bowker-Saur, London, xii + 200 pp. (ISBN 1-85739-166-7)
2. Egret, D. & Albrecht, M.A. (Eds.) 1995, Information & on-line data in astronomy, Kluwer Acad. Publ., Dordrecht, xii + 292 pp. (ISBN 0-7923-3659-3)
3. Egret, D. & Heck, A. (Eds.) 1995, Weaving the astronomy web [WAW], *Vistas in Astron.* **39**, i-x + 1-126
4. Gell-Mann, M. 1997, Lecture to ACM97, San Jose CA
5. Gibson, W. 1986, Neuromancer, Grafton, London, 318 pp. (ISBN 0-586-06645-4)
6. Gibson, W. 1993, Virtual light, Viking, London, 296 pp. (ISBN 0-670-84890-5)
7. Ginsparg, P. 1996, Winners and losers in the global research village, in *Electronic Publishing in Science*, Eds. D. Shaw & H. Moore, ICSU Press & UNESCO, Paris, 83-88 (see also the URL: http://xxx.lanl.gov/blurb/pg96unesco.html)
8. Grothkopf, U. 1997, Bits and bytes and still a lot of paper: Astronomy libraries and librarians in the age of electronic publishing, this volume
9. Hardin, J. 1993, Human collaborations technologies for the Internet, communication to *Astronomical Data Analysis Software and Systems III (ADASS III)*, Victoria BC, unpublished
10. Harman, D. 1992, Evaluation issues in information retrieval, *Information Processing & Management* **28**, 439-440
11. Heck, A. (Ed.) 1992, Desktop publishing in astronomy and space sciences [DTP], World Scientific, Singapore, xii + 240 pp. (ISBN 981-02-0915-0)
12. Heck, A. 1995, Facets and challenges of the information technology evolution, in *Information & O-Line Data in Astronomy*, Eds. D. Egret & M.A. Albrecht, Kluwer Acad. Publ., Dordrecht, 1-14
13. Heck, A. 1996, From an early electronic-publishing concept towards advanced electronic information handling, in *Electronic Publishing in Science*, Eds. D.F. Shaw & H. Moore, ICSU Press & UNESCO, Paris, 95-101 (see also the URL: http://cdsweb.u-strasbg.fr/~heck/unesco.htm)
14. Heck, A. 1997, Electronic yellow-page services: The Star*s Family as an example of diversified publishing, this volume
15. Heck, A. & Murtagh, F. (Eds.) 1993, Intelligent information retrieval: the case of astronomy and related space sciences, Kluwer Acad. Publ. Dordrecht, iv + 214 pp. (ISBN 0-7923-2295-9)
16. Heck, A. & Murtagh, F. (Eds.) 1996, Strategies and techniques of information for astronomy [STIA], *Vistas in Astron.* **40**, 361-440
17. Murtagh, F., Grothkopf, U. & Albrecht, M.A. (Eds.) 1995, Library and information services in astronomy II [LISA-II], *Vistas in Astron.* **39**, 127-286
18. Samuelson, P. 1994, Copyright's fair use doctrine and digital data, *ACM Comm.* **37-1**, 21-27
19. Samuelson, P. 1996, Intellectual property rights and the global information economy, *ACM Comm.* **39-1**, 23-28
20. Shaw, D.F. 1997, The ICSU Press programme on electronic publishing in science, this volume
21. Shaw, D.F. & Moore H. (Eds.) 1996, Electronic publishing in science, ICSU Press & UNESCO, 198 pp. (ISBN 0-930357-37-X)

THE IMPACT OF INFORMATION TECHNOLOGY AND NETWORKS: NEW PERSPECTIVES FOR SCIENTIFIC, TECHNICAL AND MEDICAL (STM) PUBLISHING

A. DE KEMP
Springer-Verlag GmbH & Co. KG
Tiergartenstraße 17
D-69121 Heidelberg, Germany
dekemp@adkathome.de

Abstract.
    This contribution can only be a small collection of ideas and experiences from my (personal) publishing point of view. The subject area assigned is very generic and there are indeed many developments going on, so I had to be very selective and restrictive, while avoiding redundancies as much as I could. I have included some metaphors and paradigms, such as the shift from print publications to electronic information. The traditional role of publishers in the information chain is compared with the new opportunities that electronic publishing is offering now and may offer in the foreseeable future. From DTP to DTD, unplugged and unbundled information, linearity, appropriateness, packaging and customizing, filters, intelligent agents, quality, integrity and authenticity are some of the items hidden in the text. The overall conclusion is: the Internet still can learn a lot from print...!

1. What publishers do!

Most scientific, technical and medical (STM) publishers are small organizations or smaller units in much larger media companies, but publishing as a whole is a big business with a long tradition and an interesting historical background. The publishers in Europe originally evolved from typesetters and printers, who imitated Gutenberg's invention, and in keeping their stock, became booksellers and later publishers. Many societies in this

Astrophysics and Space Science **247**: 11–16, 1997.
© 1997 *Kluwer Academic Publishers.*

part of the world publish in close cooperation with professional publishers, whereas in North America society publishers are more common. The American Association of Publishers (AAP) divides publishers into school, higher education, trade, professional and scholarly publishers. In this contribution I will concentrate on the last of these categories, also known as scientific, technical and medical (STM) publishers. There is an International Association of STM Publishers, registered in The Netherlands, affiliated with the International Publishers Association (IPA) in Geneva. They jointly support the International Publishers Copyright Council (IPCC). These organisations and the AAP have been instrumental in developing and adopting rules, e.g. SGML rules for manuscript preparation. One of the most important activities in 1997 is the introduction of uniform information and/or document identifiers, in addition to ISBNs (International Standard Book Number) for Books, ISSNs (International Standard Serial Number) for Serials and ISMNs (International Standard Music Number) for printed works of music. Such identifiers are critical for the identification of original works and for the success of electronic commerce on the Internet and other networks. Publishers are not at all homogeneous and the industry is not standardized, although many common rules exist. The traditional publishing process is complex: it is a complicated system with many different specialists. In the view of an increasing number of scientists as well as information specialists and in the perspective of changing information technologies and available network capabilities, publishing is costly, slow, inefficient and even a hindrance to free flow of information. So it needs to be replaced and electronic publishing should provide all the solutions...

One of the most basic activities of publishing seems to be almost unknown or simply ignored. Publishers, whether they are privately owned companies, enterprises, societies, university presses or departments, finance and administer this publication process. They sell books, journal subscriptions, translation rights and other rights and licenses, acquire advertising and sponsorships. Society publishers, controlled by their membership, may use part of their dues and offer publications as a service to their members at a more favourable rate. However, all of them have to make that difficult decision whether to publish a work or not. The decision not to publish sometimes seems to be the more difficult one. The costs of handling unsolicited and rejected material can be substantial.

2. Roles of publishers

The basic roles of publishers are concentrated in the following areas:

2.1. SELECTING, REVIEWING AND FILTERING INFORMATION

This information may be written on invitation, at the request of a colleague, editor or publisher, as a work for hire, or come in unsolicited. Journal articles fall into the category of unsolicited material, but the number of articles and/or reviews that are written on invitation should not be underestimated. Here we see a first bottleneck. It takes time to review and discuss a new intellectual and/or artistic work until the decision to invest in it and publish it, with an estimation of all the additional work that it entails, has been made. Although electronic mail may speed up the communication process, the intellectual work done by anonymous peer reviewers and inhouse editors does not change at all.

Some journals may have rejection rates far exceeding 50%. The rejection is always based on quality (relevance), the scope of a journal and its size. And not, as I heard recently from a prominent scientist, because of a lack of paper.There is some debate about how many times rejected articles are resubmitted for publication, but there are no empirical studies available. From publication experience with some major scientific handbooks in the area of chemistry and physics, I know that the redundancy in primary journal publishing is high and seems to be growing. There is no system available for filtering redundancies at the time of submission and review. Since the reviewing process is not transparent and nobody likes being rejected, there is a broad discussion of the obvious advantages of unlimited publishing on the Internet. Indeed, this social aspect of free, global and instant communication is one of the most important impacts of the Internet. But we should not forget that the Internet was developed for improved communication (and for secure information in wartime when it started as the ARPAD network) and does not simulate or even replace the publication process. I often show the (now famous) cartoon of two dogs in front of a computer: ÒOn the Internet nobody knows that you are a dog!Ê (The New Yorker). Most of the information on the Internet at the moment belongs in my opinion to communication, marketing and promotion. From the perspective of official publishing, there is too much unplugged and unbundled information, which is increasingly difficult to retrieve. Serious attempts to build databases with critically reviewed and officially published full text (written for the sake of the public record) have only been started in recent years. Most of this ÒnewÊ information is still based on print publications. Publishers select quality by screening. They add quality in the publication process, both in the printed world and in the digital age. Their buyers and subscribers expect high quality at a critical cost level. The name of a publisher and the title of a publication can work as a quality stamp. Other criteria, especially among librarians and information specialists, are continuity and availability. For electronic information, new criteria include accessiblity and retrievability. Electronic publishing is not a new activity.

Publishers and authors already concentrated on text editing and processing during the seventies and eighties (input). STM published an interesting
survey in 1995 and listed up to 120 different systems, with Word and Word-
Perfect being the most popular. The last ten years gave us new means of
output on diskettes and CD-ROMs. The near future will concentrate on the
mode of delivery: print, CD-ROM or online. Authors are not typesetters
and it is a dream to expect that all future manuscripts will be supplied in
one global format and perfect SGML coding. The successful integration of
text, images, formulae, video and sound requires new skills and teamwork
right from the beginning. The modes of delivery should be discussed as
early as possible and all information should be collected, engineered and
stored in digital archives that allow publishers to be more flexible. But it is
still a long way from DTP (desktop publishing) to the complicated world of
DTDs (document type definitions) that are essential for SGML (Standard
Generalized Markup Language).

Journals, ever since the first publications in 1665, have been and will
continue to be very efficient vehicles for the dissemination of selected content. In the future, publishers should carefully indicate what the material is
and this requires a new typology of publications and manifestations, which
has consequences for targeting and presentation activities. Springer-Verlag
has made a first attempt with its online LINK service. All bibliographic
information and abstracts have been pulled out and made freely available
as meta-information. The journal (and book) content is defined as:

- E for electronic publications (electronic sui generis)
- P for print-based publications
- S for multimedia supplements

A combination of E and P indicates an electronic version of a print
publication. In this way all those concerned with information management
and archiving know that there is still a paper archive available in libraries.
(Interestingly enough, Springer makes Annual Archive Editions of its electronic journals as a book with a CD-ROM for the database, including multimedia components).

2.2. TARGETING AND PRESENTING INFORMATION

A publisher selects and collects relevant new information for defined target
or user groups. They may be the members of a learned society, a community
of scientists (that may have started their ÒownÊ journal as a community
organizer), the subscribers of a journal, a specified group of scientists, professionals, students, or that large group of interested laymen. The content
is then prepared for better reading and understanding within that group. A
book for freshmen in medicine requires a different presentation than a reference work with mathematical formulae for physicists. Apart from vocational
requirements in the editorial process, there are also quality requirements in

the production process, e.g. better paper for medical illustrations, a larger format, different type sizes, different fonts. It is in this area that the Net still has to learn from print. In the online environment we are still heavily handicapped by the limitations of ASCII, the development and implementation stage of SGML, the Internet browsers, helpers and plug-ins. We are only at the beginning of typography. It makes little sense to present full text as a scanned image or a PDF file. Computer screens originally were designed for viewing, not for reading. For this reason we still seem to print on local printers. If there is one major paradigm shift worth mentioning, it is the transition from central printing to local or remote printing. There is one other symbolic aspect of the Internet that is not often discussed: the hypertext link! This demonstrates an important difference in presentation and a dramatic difference in cross- referring between the paper and an electronic version. I am referring to those hypertext links, usually highlithed in light blue, which are clickable. I am in favour of hypertext links as navigation tools (but they do not bring me back!) as long as they are used critically. But they might also prevent us from reading an entire article and distract us, like zapping on cable television.

In this context I should remark that a sheer transition from a linear paper-based text to an online information product is not the state of the art anymore. A parallel CD-ROM edition makes sense for archiving purposes and personal documentation. Online information should serve different purposes such as full text searching, linking, and rich indexing and offer far more functionality, beginning with interactivity and enhanced by simulation, visualization, video and sound, etc. Here we see major improvements over print. Multimedia components and supplements, 3-D presentations, annotations, moderated forums, and active or dynamic documents are all welcome additions to the typology of publications.

Electronic products and electronic versions of print products greatly support accessibility and retrievability, important barriers to the use of information in the printed world.

2.3. MAKING THE WORK KNOWN AND AVAILABLE: MARKETING, PROMOTION, DISTRIBUTION

Making the (unsolicited) work known so that it will be bought, subscribed to and cited in other works is another important task for publishers of all kinds. This is not an easy task as all those who have started a journal themselves and on their own know. Marketing, sales and distribution require considerable resources and often special skills. Here we hope to see the return on our investment within a reasonable period. Classical marketing and promotion with brochures, advertising, sample copies, and exhibitions are now greatly facilitated by means of home pages, online catalogues, news and informative texts, online ordering, useful addresses, etc. The availabil-

ity of sample pages and cover designs has supported book sales in at least
one major project (MEDOC) that I know of. Since the Internet opened to
commercial communication, publishers have used it for exactly this purpose. They still lag behind academics and professionals, who started using
the Internet for the dissemination of information and research much earlier.
In general, there is still a long way to go from classic WWW to full text
and multimedia database management systems.

All such work may be limited in the case of unsolicited articles that
are published in a journal pipeline, but copyediting, typesetting or converting to another format, font etc., proofreading, printing on paper, binding,
warehousing, packing and shipping still need to be done. Additional costs
include those for marketing, promotion, exhibitions, sample copies, and review copies. Hidden costs include trade discounts for booksellers, subscription agents, wholesalers and other middlemen who assist in transporting
the physical information product within the information chain.

3. Invisible and infinite information

The biggest problem with electronic information is not that it still requires
so many different tools or helpers: i.e. browsers, viewers, players, display
software, and sound blasters for too many different official and industry
standards. The real problem is still its totally digital and no longer physical character. In our cultures we still refer to hard copies or photocopies.
The whole process of refereeing and citing is based on the public record.
A book or journal is a hard copy of a master. Electronic documents are
something else and they can be transmitted everywhere as a clone of something that has existed or still exists somewhere. We cannot know since
no references are made. Nobody can check its originality, integrity or authenticity anymore, unless we encrypt the information. Fraud and misuse
already take place in the printed world. These developments will have a lot
of impact on the so-called information chain as a whole, up to and including archiving, preservation and conservation. Lending between libraries or
borrowing then become definitions from the past. National copyright and
intellectual property laws need to be changed and become global. How do
we handle interactive books and dynamic articles? Living documents? We
not only need new typologies, but also new rules in our global information
society. This goes beyond information technologies and the networking of
information and must be taken up as a general task for society.

ADVANCED MULTIMEDIA PUBLISHING:

WHERE DOES EUROPE STAND IN THE WIRED WORLD?

F.A. MASTRODDI
*Directorate General Telecommunications, Information Market
and Exploitation of Research (DG XIII/E)
Commission of the European Communities
Rue Alcide de Gasperi (EUFO 1268)
L-2920 Luxembourg*
Franco.Mastroddi@lux.dg13.cec.be

1. INTRODUCTION

The World-Wide Web is a good example of what we now can call "the wired world". In theory, any person at home or in the office can connect to millions of information sources, with nothing but a PC and modem. But in practice, there are wide variations in the accessibility, usability, navigability and attractiveness of digital information sources. There are several factors. The supply market is fragmented, for example there are 50,000 science publishers offering 100,000 journals. The infrastructure deployment is uneven, in some instances on a scale of one to ten across different countries. There is a rising number of digital content media: videotex, Interactive TV, Web, CD-ROM, DVD-ROM etc. Linguistic and cultural diversity also play a role. The European Union has fifteen member states, eleven official languages, and many traditions and cultures. The vast proportion of digital content is in English, for example on the Web. although over 50% of European Internet users are non-English speaking.

This paper sets out to answer three questions: what are the prospects for providers and users in Europe in the digital age? What is being done today at the level of the European Union to stimulate the process? What future plans are in store?

Astrophysics and Space Science **247**: 17–30, 1997.

2. TOWARDS PARADIGM SHIFTS

The improving digital infrastructure is inexorably leading to a number of changes in industry and society. A recent EU survey [1] gives some interesting indicators of some of these changes, which risk to become paradigm shifts in our way of working and living, modulating time and space.

In the home:

- from passive to interactive. 40 million European homes now have PCs, up from 23 million in 1990. Penetration is uneven. More Swedish families bought PCs than TVs in 1995, while fewer than 7% of households in Spain had a PC in 1994. Also, this growth is not reflected entirely in growth of on-line usage. Recent surveys in the US [2] have indicated that the penetration of Internet services in homes was only 10% in 1996. N. Doniatello estimates that the on-line market for the home is only in Year 3 of a 15-year maturity cycle.
- from narrowband to broadband. Europe's citizens are well served by conventional broadband infrastructures. In 1996, over 40 million households subscribe to cable TV. Satellite TV is received in some 40 million homes, of which 20 million have their own dishes and receivers. As these infrastructures turn digital, an enormous capacity will become available for interactive information services.

At work:

- Over 72% of office workers in Europe have a PC or equivalent in their workplaces. New knowledge acquisition will become a way of life. According to the US Department of Commerce, by the next century, 60% of the new jobs will require the kind of information and computing skills that only 22% of the workers have today [3].
- from the fixed office to the virtual workplace. There are an estimated 1.25 million teleworkers in the EU, in a working population of 140 million people. There is still a small percentage, but growing rapidly over the past two or three years.

In the publishing world:

- from linear (print) to non-linear (random access databases and archives)
- from isolated researching/authoring to collaborative tele-work
- from permanence (paper and ink) to the ephemeral (indefinite number of digital versions)
- from one-dimensional presentation (e.g. flat presentation of ASCII or WYSIWYG data) to multi-dimensional ones (2-D, 3-D and animation such as living books)

– from mass marketing (e.g. journals and newspapers) to do-it-yourself publishing and individual customisation (e.g. personal information delivery).

3. THE ELECTRONIC PUBLISHING INDUSTRY

Concerns have been raised about the impact of electronic publishing over traditional print publishing. Is print publishing going to disappear? Will new technologies disrupt several hundred years of science publishing? There is no definitive answers to these questions. However, recent surveys and observations have injected some clarity into the picture.

The current strength of the print publishing industry is uncontested. A recent survey by Marshall Inc. [4] reckons that the print publishing sector in the US accounts for some 100 billion US$ revenues per year. This compares favourably with Cable TV (29 B$), film entertainment (20 B$) and pre-packaged software (17 B$). In Europe, the overall turnover of the content-related industries is higher than telecommunications or information technology. Europe has over 2200 daily newspapers with a circulation of 194 million, representing a penetration per head of population higher than the Americas and Asia combined. Europe also has ninety-six thousand libraries with over one billion volumes available to the public on a daily basis.

Electronic publishing is expected to take up a significant portion of the print publishing market by the year 2000. The penetration rates are expected to differ according to the different sectors, as the table below illustrates, for Europe. Science publishing is expected to be a particularly fast growth area, and this has been backed up by recent announcements by large publishers like Elsevier and Springer to invest heavily in electronic journals. Despite the positive trends, take-up is slower in Europe than USA. By May 1996 more than 1100 newspapers world-wide had on-line editions [5]. More than half of these are US-based. Only one quarter are European.

Where are the markets in Europe?

A survey of the professional on-line market by IIE, Cologne [6], based on 1994 figures, shows that the main markets for electronic publishing are, not surprisingly, in the UK (28%), France (18.5%) and Germany (14.4%), in terms of user expenditure. The Netherlands is active beyond its size with 6.3% and the Scandinavian countries total some 14%. Whilst these figures will have evolved, the overall balance of expenditures has probably not moved significantly. Internet services generate much traffic (100 million hits per day on Netscape), but little user expenditure.

The consumer market is booming.

TABLE 1. European Publishing Market segments 1996 and 2002 and their potential for Electronic publishing

	1996		2002	
	EP Market Volume (MECU)	% of total market	EP Market potential (MECU)	% of total market
Newspapers	310	1-2% (1)	2700	3-6%
Magazines	200	1-2% (1)	2000	3-6%
Book market (total)	250	0.5-2% (1)	1700	3-5%
– STM/specialist	50	3-5% (1)	600	12-15%
– Educational	70	1-3% (1)	420	8-9%
Corporate	375	2-5% (2)	2400	10-15%

(1) in terms of revenues
(2) in terms of copies produced
(3) including catalogues & directories
Source: Andersen Consulting/IENM [5] (medium scenario given)

1995 was the most fruitful year yet for consumer on-line services, as nearly five million US households opened accounts, bringing the end of year total to 11,304,200 customers, a staggering 79% increase over the 6,320,650 households that were wired at the end of 1994, according to the latest quarterly census of 22 consumer on-line services by *Information & Interactive Services Report* [7]. If this growth rate remains, there could be 170 million on-line users world-wide by the year 2000.

The six largest national/global on-line services now represent 96.9% of the customer base. The three largest operators (America On-line, CompuServe and Prodigy) alone reach 10.1 million users, or 89.3% of the total audience. The other three services (Microsoft Network, eWorld and Delphi) account for some 0.8 million users.

However, this growth rate has not been without problems. There is a notorious amount of ôchurnö, namely abandoned user accounts. Saturation is setting in, as the 1 million hours per day benchmark is being reached, provoking network crashes, slow processing and user complaints.

Is CD-ROM being left behind?

The current fad with the Web, combined with some market announcements about downturns in investments could give the impression that the CD-ROM market is on the decline. This is not (yet) so. In 1996, the world-wide CD-ROM installed base passed the 100 million mark, with the US,

Japan, Germany, Canada and the UK leading the way. Total expenditures exceeded 30 billion US$ world-wide [8].

In fact, CD-ROM has found a new lease of life, acting as a carrier for client software and plug-in applications, and helping to spawn hybrid on-line/off-line services. It has become an integral part of the multimedia PC.

Is language diversity a problem?

A Web user survey by GATECH [9] recently quoted that language is not an issue on Internet, but immediately qualified this by adding that the main population asked were American students! European users were more concerned, naturally. In science publishing, the nearly universal language is English. 92% of STM databases have English, compared with 88% in 1995. Other languages covered are French in 5% of cases (4% in 1995), German with 4.5% (3% in 1995), Spanish with 3% (2% in 1995) and other languages with 3% (same in 1995).

This factor is probably not seen as a major hindrance to the increasingly international academic and research communities in STM, but could become a factor if electronic publishers wish to target broader international markets or localise their products. Two-thirds of the STM databases surveyed originate in the US or Canada. Just under one-quarter (one-third in 1995) originate from within the EU member countries, whilst a small but growing percentage comes from Australasia, East European or other countries.

However, in other areas, especially those closer to the office worker or citizen, the language barriers are real. In 1995, the UK's DTI reported that 30% of UK export companies were losing business due to the language barrier. The problems for continental Europeans, as well as for Asians and other parts of the world, would be bigger. It is no coincidence that Japanese industry has invested heavily in machine translation systems.

Today, over 50% of European Internet users are non-native English speakers. This percentage is expected to grow substantially over the next decade. This could explain why some consumer and commercial operations are beginning to integrate language technologies into their systems. For example, CompuServe plans to add facilities to help users scan foreign language information sources through machine translation. Reuters is providing more sophisticated news selection and delivery facilities based on pre-set user preferences, and content-based filtering of newswire dispatches.

Who will manage the information explosion?

Science publishing produces some 1.5 million works (articles, papers, monographs, etc.) per year. The accumulating scientific and cultural heritage needs to be stored in future in a way which allows quick and easy access. Yet, libraries, museums, galleries, archives and other organisations which preserve and give access to our heritage are currently fighting to keep up with the wired world.

The Clinton-Gore Administration has announced its intention to help link every American library to the Internet. In Europe, there are 96,000 libraries, but only a handful are networked. Today, there are perhaps 250 libraries around Europe which have Web sites, compared with 800 in the USA (including 475 in the academic area and 200 public libraries). The rest of the world accounts for some 130 sites (50 in Australasia, 60 in Canada, 20 in South America). The potential in Europe is growing. The British and French national libraries have made substantial investments in digital content technology. The Dutch on-line library network PICA is one of the most advanced in the world. The children's Internet library project called CHILIAS, sponsored by the European Commission Libraries Programme and the Stuttgart Public Library, is a forward-looking example of what can be done.

In the museums area, it is a similar scene. The Louvre has just released a CD-ROM with the well-known pictures and works of art. Yet, over 80% of the museum's holdings are still in original form, stored away from the public, due to lack of shelf space. It is the same story for many other big museums. They have recognised this problem, and for example 400 museums and art galleries have linked up through a Memorandum of Understanding on multi-media access development over the next years.

Archives present a difficult problem. A 1994 report of the European Commission on archives asks the question: are we to witness a loss of memory in computerized archives? There are new problems pertaining to computer law, the rapid technology developments and the sheer effort needed, which risk to impede the transition of valuable public or private archives into the digital age [10].

3.1. THE SUPPLY SIDE: WEB EXPLOSION

3.1.1. *Internet-based services*

Out of all the services springing up on the Internet, probably the most attractive ones in terms of publishing are on the World-Wide Web (WWW). The Netcraft survey collects and collates as many host names providing an http service as it can find, and systematically polls each one with an HTTP request for the server name. In the February 1997 survey they received responses from 739,706 sites (up from 430,000 in October 1997) [11]. It must be said that other similar surveys come up with different figures, relating to their definitions and coverage.

What are the perceived advantages? Speed is a big advantage seen by publisher and reader, by expediting peer review, editorial and production work, and distribution. Improved access to information and hyperlinks to other electronic information, such as between reference citations and article abstracts, between references etc. Publication of new multimedia objects (still images but of animation, movies, and sound) and digital archives e.g. on CD-ROM are other benefits. It can enhance and encourage more scientific discourse about research by allowing "Letters to the Editor" and discussions. Are these services attracting users? The British Medical Association logged 10,000 on-line searches in one month on its recent dial-up/telnet Medline service.

Science publishing has exploded on the Internet in the past 2-3 years. This is maybe not surprising as and academic and scholarly publishers in Europe were amongst the first to recognise the potential importance of the Internet as a publishing medium. Today, there are approximately 20,000 science site entries listed on the Web by Yahoo (February 1997), compared with some 4,000 in 1995. Astronomy, for example, now rates over one thousand entries. Many of these sites are for contact and reader services, and most do not contain full texts. Table 2 gives the evolution in selected topics.

TABLE 2. Evolution of Yahoo entries in selected topics

Subject	1995 entries	1997 entries
Physics	469	1073
Computer Science	727	1445
Earth Sciences	473	1984
Biology	509	1989
Engineering	880	3124
Medicine	612	806

3.1.2. *Electronic journals on the increase*

Amongst the many science sites, there is a fast-growing number of electronic journals. Edoc lists 350 peer reviewed journals on-line. The vast majority of journals (207) are on US sites, although some are European journals. A significant number are hosted on sites in the UK (35), and Germany (25), whilst the rest are sprinkled across France, Eastern Europe and Australasia [12].

The main European science publishers like Elsevier Science, Springer-Verlag and Oxford University Press are active on the World-Wide Web.

Elsevier Science, for example, is following an evolutionary path towards becoming fully on-line, if the future market warrants. They have set up an on-line catalogue of their published works. In the future, each entry will have its own home page, links to other sites then access to an on-line version. They already offer some on-line products. For example New Astronomy offers a home page and an article snapshot with abstract, scaleable drawings (from thumbnail to full size) and references. The on-line presentation has some visible limitations, including problems with outsize characters, sub- and superscripts and grey scales. These limitations are overcome somewhat by offering an alternative house-style format in PDF or PostScript.

The Springer site offers over twenty on-line journals through subscription. Oxford University Press provides a home page for all of its 160 journals, with abstracts and reader information. There are several sample electronic issues, some of which deliver full text, two weeks ahead of the print publication.

Who can provide the essential indexing and marketing skills? An interesting initiative by OCLC (Online Computer Library Center Inc.) to provide a platform for electronic journals, makes clear the potential contribution of library services to publishing. Not only does OCLC represent a substantial market (20,000 libraries are linked to their network), but also they can provide information indexing and classification skills. They currently provide 48 on-line journals through Internet and dial-up, and are developing Web interfaces to these.

What does the future offer? There are strong sentiments that the Internet publishing will not remain viable without key factors such as: quality control, data integrity, an open and trusted transaction scheme, better navigation and retrieval, and easier linguistic access.

3.1.3. *Conventional STM databases*
Both the number and variety of electronic information services in the STM area have seen continuous expansion in the last twenty years, but the core supply is still represented by classic on-line bibliographic databases. In total, approximately 1300 (1000 in 1995) of the world's 9000 or so on-line databases are listed as covering the science area, although the figure is probably nearer to 900-1000, allowing for different versions of the same databases and sub-files of a database series [13].

CD-ROM format is showing a slight growth. 48% of STM databases are on-line (down from 54% in 1995) and 31% are on CD-ROM (up from 23% in 1995). Other formats are diskette, magnetic tape or handheld.

The professional databases are still mainly bibliographic. Over 53% (40% in 1995) of the databases are bibliographic in format. A growing number (36% up from 29% in 1995) contain full text. Only a small but rising percentage of the databases carry images: 8% as opposed to 4% in 1995.

3.2. THE DEMAND SIDE: NEW USER MODELS

There are many barriers to determining user behaviour in the field of electronic publishing. It is a young activity, fragmented across many different disciplines, income brackets, geographic areas and socio-economic profiles.

The GVU surveys show that over half of Web users are in the 16-25 age bracket, only 5% would pay fees, the vast majority are male and spend more than 20 hours a week on-line. This bias is probably due to the survey being based around universities. However, it illustrates that Internet demographics can be very different from professional on-line, CD-ROM or television.

Information providers, whether commercial, institutional or individual, lack information needed for setting up business models, quality levels, user support, etc.

There are strong indications that a new way of describing user behaviour is needed. Whether this is according to attitude ("the enthusiast", "the oldliner" etc.), proximity to the content (intermediary or end-user) or economic factors is highly debatable. The recent Andersen Consulting study [5] for the European Commission made an interesting contribution. It divides users into four main segments:

- knowledge workers who "must have" information for their work, through any means. An extreme example is researchers in the pharmaceutical industries, who can sometimes spend more on information retrieval of previous work than on research activities;
- PC-enthusiasts who are the early adopters of complicated new media and information technologies. But will this group disappear if and when simple information appliances hit the market?
- time-constrained users, mostly professionals who depend heavily on premium information, but have no time to look for it! Examples are executives, lawyers and doctors;
- leisure seekers, who use information in their spare time for entertainment, self-education and reference. This section is also described as the "plug-and-play" segment of the market.

This type of segmentation, like many others, carries the seeds of a new model of user behaviour, but still has to stand the test of time. The critical factor is whether publishers and providers can base business models on these user segmentations.

How user-friendly is the Web?

The Internet greatest asset, according to the Economist [14] is also its greatest liability. It lets anybody publish world-wide, at little or no marginal cost. This has led to some chaos, however, only partially remedied by directory services like Yahoo. Web pages are random, unstructured and of variable quality. The most that current search engines can offer is free text searching. This inevitably produces misses and false hits. Boolean logic cannot help if there is bad indexing and poor classification. The problem is even worse in non-English language, as many Web search engines are geared for the English language and grammar.

There are probably 80-100 recognised search engines easily accessible on the Web, such as Lycos, FIDO, Webcrawler and Alta Vista. They have many different characteristics, such as being customisable, domain-dependent, geography or language-dependent, more or less sophisticated, more or less ôintelligent,ö etc. Their capabilities range from simple browsing based on one-word searches to more advanced searches allowing Boolean and proximity searching and results ranking. However, they are dependent on how well the Web site is indexed, which is not as extensive as with classic databases. Also, a Web search often does not give detailed information about a site's contents, but a title or sometimes just a journal issue number. In such cases, the results of a search often need to be verified by downloading the full text, with all the potential waste of time and energy.

None of the current Web tools allow to search on the content of images or video clips. Documents, still images and video, as well as sound files can nowadays be located and downloaded relatively easily. The problem is in searching the contents of the multimedia object, for example to locate a certain detail, shape or pattern, to be able to interpret molecular structures in a visual way, or to extract data from medical images. The objects would need to be tagged with 'meta-information', and the search engine would need to recognise this.

4. EU-FUNDED ACTIVITIES

Most EU actions relating to the future of electronic publishing are in the research programmes. The current *Framework Programme for Research and Technological Development* (known as FP4) runs from 1994 to 1998.

4.1. CURRENT RTD ACTIVITIES

Today, there are several EU-funded research and market stimulation initiatives which cover different aspects of electronic publishing. The most relevant of these are the new research activities in information engineering, libraries and language engineering.

These are research activities inside the *Telematics Applications* programme under FP4, which covers applications of information and communications technologies in areas of general interest like health care, transport and education.

4.1.1. *Information engineering research*

Information engineering is a relatively new term for an old business, namely information processing, packaging and use. It is a new research area with a current budget of 37 MECU. Its aim is to permit a more selective access to information in all its forms through the application of telematics-based methods and systems, focusing on information content.

Between 1996 and 1998, the sector supports nine pilot applications, ten feasibility projects and a number of support activities e.g. on standards and concertation. In the field of electronic publishing, for example, Multimedia Broker develops new tools for innovative multimedia products. MBLN covers local newspaper publishing and advertising. Europe-MMM develops tailor-made packages of learning materials. Twenty-One explores the problems of large and complex documents with multimedia elements. Two more detailed examples are given below:

GEOMED. The growing acceptance of geographic information system technology, and its role in the wider trend toward collaborative working within and between diverse organizations, poses some practical challenges. The GEOMED project seeks ways to integrate services for creating, disseminating and using geographical information, specifically in the sectors of urban planning, public administration and environmental protection. It addresses the practical integration of geographic information with other forms of documentation. It also looks at inter-operability among proprietary GIS products. This involves issues such as converting proprietary GIS data into standard formats for exchange between software products and hardware systems of various kinds, as well as for use in electronic mail systems, videoconferencing services and hypermedia environments such as the World Wide Web. The object here is, obviously, not to develop yet more products, but rather to facilitate the exchange of information and services between different platforms and packages.

The MAID project seeks to improve communication and information services within the industrial design sector with a range of facilities including a multimedia database, design tools and utilities, together with facilities for collaborative working between companies regardless of their location. The Information Centre would provide resources including on-line access to industry standards, research papers, product literature and the current catalogues of participating organizations, so that designers could simply search for the data, products or suppliers which they need at the moment, rather than struggle to maintain their own reference collections. The Service Centre could include access to design tools and related software products (including, for example, workflow management systems for the administration of collaborative projects and shared workspaces). The Centres will eventually provide commercial services, with appropriate charging mechanisms. A consortium of 25 organizations from nine countries is coordinated by the *Centre for International Technology and Education* (CITE).

4.1.2. *Telematics for Libraries*
With the advent of the Information Society, there has been an explosion in resources, generating new demands and fuelling higher expectations of libraries. Technological advance has created a new world in which information is produced, accessed and used in electronic form. *Telematics for Libraries* aims to improve and expand networked library services throughout Europe as well as network connections to couple library resources with external assets, providing seamless access to published information from a variety of sources.

Alliances between publishers, distributors and librarians will aim to provide new levels of service, accessible from the library itself or directly from homes, schools and offices. The programme currently runs some 22 pilot application projects under FP4.

Some of the major issues tackled by the libraries programme concern:

- public libraries, where a number of projects address community information, services for adult learners, childrenÆs learning and distance learning,
- European national libraries, focusing initially on national bibliographic records and international catalogue exchanges, and now addressing issues of record format, name authority records and statistical data
- library networking, especially for cooperation and resource sharing.

4.2. FUTURE EU REQUIREMENTS AND OPTIONS

4.2.1. *The need to add value*
The main requirement at EU level is to add value to the existing trends and developments in the marketplace and in the public sector. Added value is in creating fresh momentum, by looking forward to the next years, and not by playing 'catch-up'. This includes the stimulation of new industrial and service-sector bases in digital content, anticipating new European require-

ments for language and cultural components in ICT systems, promoting scientific, cultural and educational exchange, creating collaborative teams in new, interdisciplinary areas, taking on board the needs of European users for user-friendliness, information handling and access functions, etc.

The added value in terms of research objectives could come in two complementary areas. Firstly, more creativity and advanced design of information content is needed to stimulate a whole variety of new multimedia information services, whether film, data, text, voice, etc. As Sir David Puttnam said in a recent EU seminar, the current multimedia products are like ôthe old black and white films of the early days of moviesö. Secondly advances like speech interfaces, navigation, searching and transactional services are needed, to stimulate the use and usability of information.

Added value will also come in the new synergies to be created. Electronic publishing drives convergence between technologies, by combining content with communications and communities. It is cross-media, bridging the gap between audio-visual and computing.

4.2.2. *What are the options?*

The main approach for future RTD actions, in the fifth Framework Programme 1999-2002, as set out by the European Commission and many consultative bodies, is to increase the impact of research on competitiveness and employment. The FP5 proposal suggests to concentrate RTD around a limited number of main themes. One such theme revolves around the Information Society and its growth in the next decade. Better forms of digital content, easier accessibility and usability are key factors within the Information Society, developing in parallel with. and not behind, the evolving technology base and infrastructure.

Interactive electronic publishing is one priority topic, as it can tackle the whole new value chain of publishing and information handling, independent of specific technology options. The cultural content side should also be covered, for example through digital libraries or virtual museums.

New language technologies, whether for monolingual or multilingual applications, will help to make information and communication systems more user-friendly for Europeans.

Finally, technology advances in information access, filtering and analysis should be stimulated, e.g. intelligent agents, intuitive and immersive interfaces etc. New tools and methods will help overcome the information explosion and give normal citizens increased trust for commercial transactions, reliability and protection e.g. liability and anonymity.

References

1. The Information Society and the citizen. A status report. European Commission Sept. 1996 – URL: http://www.ispo.cec.be
2. Doniatello, N. 1997, Odyssey Market Research, MILIA conference proceedings, France – URL: http://www.reedmidem.milia.com
3. Remarks of Commissioner Susan Ness before the American Library Association, Washington, 16 February 1997
4. Marshall Marketing and Comunications Inc. 1997 – URL: http://www.mm-c.com/
5. Strategic developments for the European publishing industry towards the year 2000. European Commission, DG XIII/E, Luxembourg 1996 – URL: http://www.echo.lu
6. Bredemeier, W. Market development of Electronic Publishing in the Member States, IIE, Cologne, Germany
7. IISR Report Jan 1997 – URL: http://brp.com/netline/iisrnews.html
8. Schwerin, J. 1997, Special Report ôThe winter o our disc content, InfoTech, Woodstock, Vermont
9. Graphics, Visualization and Usability Centre, Jan 1997 – URL: http://gatech.edu/pitkow/survey/
10. Archives in the European Union. European Commission. OPOCE, Luxembourg, 1994, ISBN 92-826-8233-1
11. Netcraft, Feb 1997 – URL: http://www.netcraft.co.uk/survey/
12. http://www.edoc.com/ejournal, Feb. 1997
13. Gale Research Databse, Datastar online service, Feb 1997.
14. The Economist, Sept. 14th 1996, p.12 The Total Librarian.

THE TECHNOLOGIES WHICH UNDERPIN
ELECTRONIC PUBLISHING

F.D. MURTAGH
Faculty of Informatics
University of Ulster
Magee College
Londonderry BT48 7JL, Northern Ireland
fd.murtagh@ulst.ac.uk
http://www.infm.ulst.ac.uk/~fionn

Abstract. Astronomy as a discipline is essentially online and digital. It thus offers an ideal framework for the examination of current trends in electronic publishing. Electronic publishing is driven on the winds of various underlying technologies. We look at a number of these, focusing on information access and retrieval.

1. Introduction

Underlying methods and techniques play a major role in our professional activities, and how we carry out these activites. This is true, too, for electronic publishing. The coming to prominence of the World-Wide Web from early 1993 (with the release of the Mosaic browser) is a dramatic illustration of this. Whether authors or publishers, our lives are irrevocably changed by the events of early 1993.

Important steps forward, such as the Web, bring with them further issues and requirements. We will look at some of these in this chapter. We adopt the perspective of publishing in the areas of astronomy and physics being increasingly online. We jump, then, to some of the issues which are on the research and development agenda as a result of this. The treatment of topics covered is selective, but it is attempted to overview some of the potentially most important topics.

The technical and methodological problems posed by electronic publishing include the following.

Astrophysics and Space Science **247**: 31–40, 1997.

Increasing volumes of online data and information imply the need for suitable naming conventions. Ultimately, these have to permit natural language querying of multimedia (data on different media), multimodal (the same information in different guises – sound and text, for instance), multilingual, distributed, heterogeneous data stores. Such data characterizes what the scientific and technical publication is. Motivation for the scientific publication from the 17th century has been (i) human communication, and (ii) professional brownie-points.

A phase beyond naming conventions is that of content-based search. We will look at directions currently followed in academic information retrieval. Search and access are closely related. An issue in the latter area is that of the user interface. Older command-line user interfaces, especially over a slow computer communications link, are unpleasant to use. We will look at recent work incorporating spatial metaphors into the access and search process.

Next, we move on to the issue of bandwidth and network response. This is an area of incredible pain and agony on the part of the user. Bandwidth must be bettered, but there is another approach which ought to be employed in parallel to a greater extent. This is information filtering and compression. The simple fact is that noise, by definition uncompressible, occupies the very major part of all current information and data flows.

Finally, we will take a brief look at where these technical and methodological issues are taking us.

2. Resource Description

In this section, we look at how access to information is facilitated through resource (text, documents, data) description.

We assume that the information we need to search through, or which we need to allow users we support to search through, is on the Web. In fact, the all-pervasiveness of the Web for production and distribution of scientific information could lead us to characterize all non-Web areas of life as (perhaps unforgiveably badly formatted) Intranets!

A priori naming conventions range from the Dewey Decimal Classification system used in libraries to the bibcodes used in the Astrophysical Data System (ADS) [1] to uniquely characterize each document. Bibcodes cover some chosen disciplines. They do not cater for document content.

Tagging of content is a step further towards facilitating content-based retrieval. Data can be structured in quite diverse forms though – HTML, SGML, X.500, SQL databases, email files, EDI (electronic data interchange) and so on. One standard for environmental information which seeks to be above all of that is the Global Information Locator Service (GILS) [2], originating in US federal global change environmental work. The server conformant to GILS supports searching of the characteristics of information, at varying levels of aggregation. This is achieved through the availability of hand-crafted or automatically generated "locator records". An example of these is the descriptive information needed to support spatial querying, based on latitude and longitude. Various commercial and public domain servers offer GILS compliancy [2], e.g. SIRSI Vizion, Isite, Verity, BSN (Basis Systeme Netzwerk), OLCL (Online Computer Library Center), and Fulcrum Surfboard. In addition servers such as AOLserver, Netscape Commerce Server, and Apache freeware server, offer support.

Parallel work on library information is described in [3]. Information access and resource discovery, it is stated, is increasingly a problem of effective management of metadata. Metadata is data describing data. More precisely, support is provided for location, level of precision, origin and history of processing, selection, etc. Differing degrees of functionality are offered in differing disciplines – social science, geospatial, museum information, and other fields. Bibliographic metadata standards include MARC (machine-readable cataloging) and Bib-1.

A system of this sort in astronomy, called the Astronomical Server URL [4], is premised on the fundamental astronomical data object known as a catalogue. The URL (uniform resource locator, or Web address) is extended, by means of a CGI program running at the server, to handle

```
field1=value1&field2=value2...
```

The acceptable fields (metadata) are covered in the standard. For relational databases, SQL format is also accepted.

A brief historical overview of query and associated indexing standards, in particular for public domain software packages, was given in [5]. An extensive description of such packages is included in [6].

Among systems which are available are freeWAIS-sf from the University of Dortmund [7]. Based on the WAIS (wide-area information system, Z39.50–v1 or Z39.50–1988) standard, it modified and enhanced the original WAIS software. WAIS Inc., spun off by the parent of WAIS, Thinking Machines Inc., enjoyed an unsettled childhood, being sold to AOL in 1995, and subsequently to Fulcrum Technologies.

Isite from the Center for Networked Information Discovery and Retrieval (CNIDR) is based on the Z39.50 (Z39.50–v2 or Z39.50–1992) standard.

GAIS (global area information system) [9] is software from the National Chung Cheng University of Taiwan.

In all cases, gateways are provided to allow use on the Web, i.e. to allow access to the HTTP server daemon. Fig. 1 illustrates how the Web (the HTTP server) provides a convenient front-end to various data stores. Data selection and data processing can be carried out before returning an HTML-formatted output for passing on to the client browser.

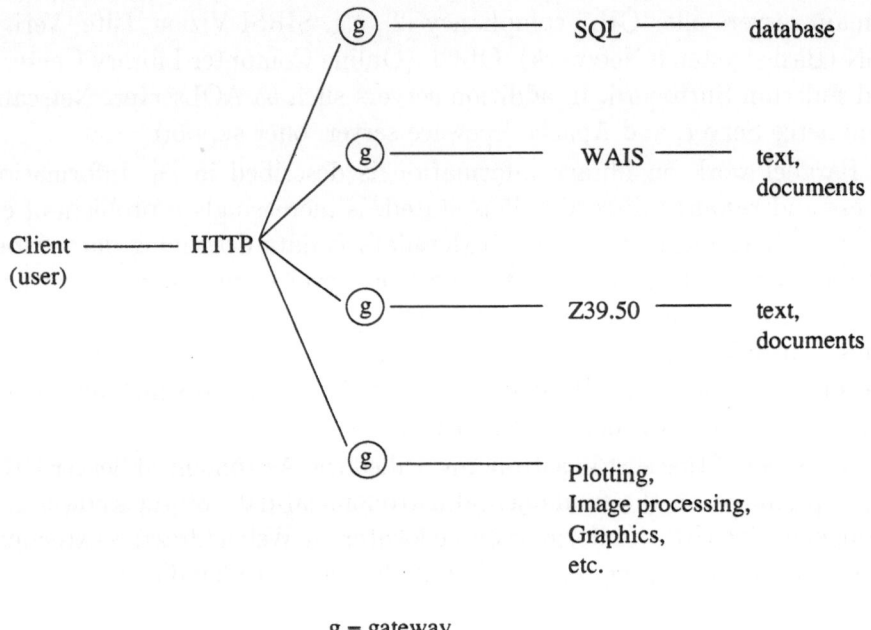

g = gateway

Figure 1. Client-server architecture. Web interface to relational databases, statistical data collection, free text, and so on.

Such indexing and search systems may differ in their functionality, including (see [6]): gateways to HTTP; relevance ranking algorithms; use of stop lists, synonym lists and thesauri; right and left wildcarding; support for fields; support for numeric data, e.g. date ranges; relevance feedback; document format support, e.g. PDF, MS Word, etc.; soundex and phonix; tolerance of misspellings; and so on.

3. Searching using Concepts and Content

In this section, we look at attempts to characaterize concepts; and proceed to content-based search.

To advance towards concept-based searching, standard information retrieval approaches must be scaled up [10]. In short, this direction takes us to concept spaces based on index terms. We will briefly look at where traditional information retrieval is currently headed. Then we will sketch out the highly important area of Web-based information retrieval. In passing we will note how fundamental the query is (i.e. what the user wants), which motivates the final part of this section dealing with interactive user interfaces to textual and other information.

'Classical' information retrieval has not perhaps achieved any quantum leaps in the last decades. Nonetheless, older results have stood the test of time: that recall and precision are still used for system appraisal; that indexing and inverted files cannot be avoided; and the mechanism for query expansion, known as relevance feedback (adding results of interest to the user following a first iteration query to the second iteration query) is important for quality of results. For a very useful compilation of papers and bibliographies, see [11].

An interesting departure in recent years is that the small artificial document test-sets, given names associated with their originating institutes or organisations such as INSPEC, CACM and NPL, long used by the research community, have been replaced by the annual information retrieval competition TREC (Text Retrieval Conference) [12]. TREC-1 was held in 1992. For each conference, textual data is available, with queries and (most importantly) relevance judgements. Texts, available for testing and appraisal in advance, are currently moving from about 2 GBytes to about 20 GBytes in size.

In the context of analysis of such test data collections, an interesting point brought out in [13] relates to query size, i.e. the average number of terms used in a query. In this work, it is found that a particular processing strategy is most effective when queries are large in size (and are used in a static environment). Long queries may be expected with relevance feedback. They are also used when querying ADS [1] with document abstracts or even entire papers as the query. Catering for long queries is thus an important objective. It is also diametrically opposite to Web searching, which we will look at next.

Web-wide searching represents presumably the dominant information retrieval practice today. Queries are often very short, consisting of 1 or 2 terms. Search tools are quite primitive. Recall, and efficiency, are prioritized, and precision is sacrificed. Web search engines use man-made summaries, or automatically harvested information. Web documents are partially or fully indexed.

Manually-organized lists include very widely-used discipline-specific services in astronomy and physics [15].

In harvesting, a list of URLs is taken, and depth-first or breadth-first traversal strategies implemented based on links in the given documents.

The Web currently contains over 100 GBytes of textual information alone [14], or about 20 million identified documents. Clearly data from databases, image data, and so on are not directly included in these figures. Harvesting of textual information in its own right is non-trivial. Among proposals and work for improvement in Web-based searching are more adaptive approaches, including mechanisms based on hierarchical indexing (sometimes termed centroids) and self-regulating hyperlink-based systems (see discussion of Ingrid below).

The classical vector space model of information retrieval is inherently spatial. Humans seem to have excellent locator and discovery mechanisms which are based on spatial relations. It is not surprising therefore that spatial metaphors are widely used in information retrieval. Apart from user orientation, user interfaces based on these principles may be of help in improving user interaction.

The combined clustering, spatial projection and display method known as the Kohonen self-organizing feature map is currently being studied for its advantages and interesting properties. Fig. 2 shows a summary of titles, abstracts and keywords from *Astronomy and Astrophysics* between 1994 and 1996. The map shown is content-addressable and gives access to lists of grouped documents, and their characterizing keywords. Currently at an early prototyping stage, the system is already integrated with ADS. Not surprisingly, this type of interactive user interface has excited widespread interest [16]. For further details of the work underway in astronomy, see [17].

4. Filters and Abstracters

The following section explores the saturation of bandwidth which is so typical of present Web usage.

So far, we have had in mind the quality of the searching or querying result obtained by the user. Another aspect is also of importance. This is the infrastructure, and data volume, associated with an increasingly online society. The interesting point is made in [18] that the Web came about when the Internet's bandwidth surpassed a certain data rate. The backbone of Internet 2.0 operating at 45–155 Mbps allowed multimedia traffic to be carried, and subsequently to the light and effective Web system based on the Mosaic and later browsers. Bandwidth is supremely important. Without it, we grind to a standstill. And we are grinding to a standstill, in particular in Europe. International connections are often the big problem. A regularly updated map showing connection infrastructure which holds for much of Europe is available at [19].

While we choke from lack of bandwidth, and while this sets unsurmountable limits to further progress (in teleconferencing, or transmission of large

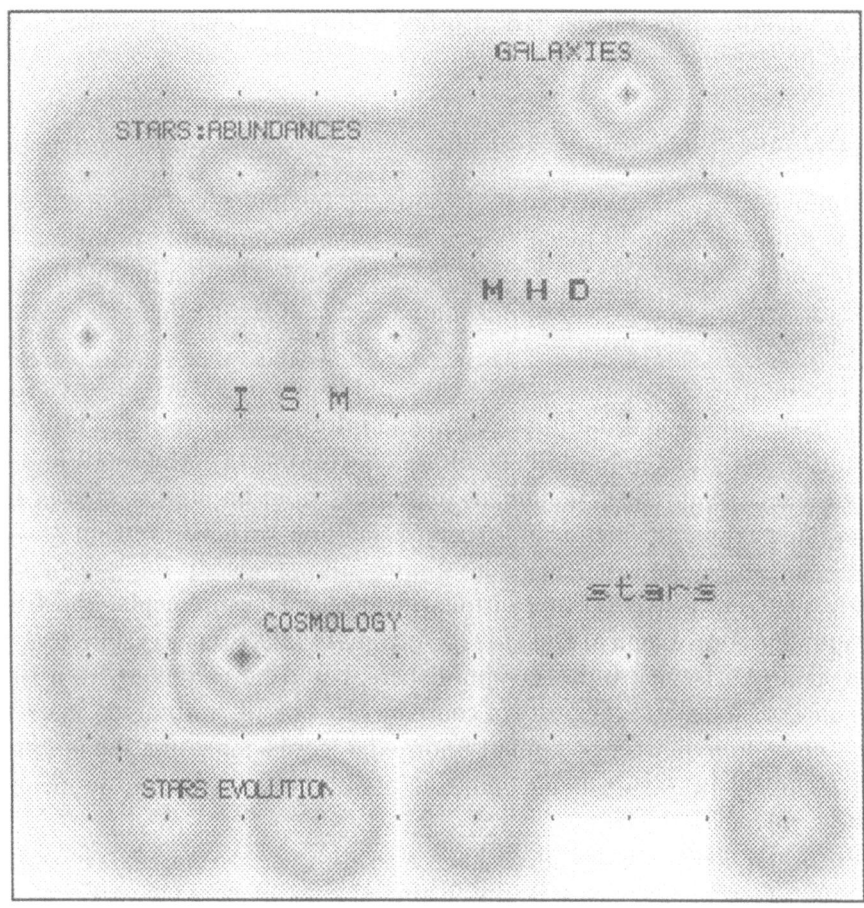

Figure 2. A cartographic and interactive interface to *Astronomy and Astrophysics* articles in 1994–1996. Lists of documents are associated with nodes. Colour indicates relative density of documents. Some characterizations of nodes are provided.

data files, for instance), we should pursue approaches to text summarizing and text denoising. The problem of image transmission is enlightening. Scientific data consists to a large extent of noise. Denoising with insignificant damage to the data [20] is a genuine service to the user community. Building filtering into communications protocols in discussed in [21]. At a higher level, automated summarization of texts and documents is a field which is still at a nascent stage [22].

There is an urgent need for much more of this work, – for automatic finders of astronomical object names in texts, for automatic finding and summarizing (even if partial) of numeric information in documents, for automated hyperlink creation and handling of citations and references. The latter includes links to image and other databases, some of which is currently in place and operational.

5. Balance-Sheet and Trends

In summary, we find that many of the basic technologies to support electronic publishing are in place. There are however mounting problems due to the success of the Web. Solutions to these problems often imply strategic choices. Automation provides, in some measure, a key to these choices and solutions.

The increasing volume, alone, of the published literature requires solutions in the direction of summarization and information filtering. Automatic document matching potentially facilitates the finding of enthusiastic and expert referees. Automatic document matching could be used to closely link different versions of documents, thereby helping with the version control problem. Object and person name finding allows speedy integration into services such as SIMBAD (Set of Identifications, Measurements, and Bibliography for Astronomical Data) [23]. Automatic alert services, currently in operation at Elsevier and other publishers, could be enhanced from titles and authors to selected aspects of content. Most of all, automated insertion into the web or matrix of human knowledge could make for the scientific and technical literature being more accessible and ultimately more used than is currently the case.

In other fields, similar ideas have already been proposed. The Ingrid information grid [24, 25] sees publishing taking place when a document publisher "announces" the availability of a document to Web servers that it knows contains similar documents. A server receiving the announcement may choose to create links from similar documents of its own to the new document. Over time, documents on a given subject become linked together, building up a mesh topology, and making it easier to browse and access information. "The act of generating these [descriptive] terms", it is stated in [25], "and of inserting the resulting Ingrid node into the Ingrid topology is called *publishing*". It is pointed out how, in the extreme, a resource that has little intrinsic value on its own may become valuable in the Ingrid topology. Knowledge, or stored and collective information, is the topology. This is a nice idea which will certainly appeal to the author whose primary wish is for immediate open-access availability of his or her output.

References

1. "The NASA Astrophysics Data System", http://adswww.harvard.edu; "Abstract

service bibcode query help", http://adsdoc.harvard.edu/abs_doc/bib_help.html

2. E.J. Christian, "GILS. What is it? Where's it going", D-Lib Magazine, December 1996. http://www.dlib.org/dlib/december96/12christian.html

3. L. Dempsey, "ROADS to Desire. Some UK and other European metadata and resource discovery projects", D-Lib Magazine, July/August 1996. http://www.dlib.org/dlib/july96/07dempsey.html

4. M. Albrecht et al., "Astronomical server URL", version 1.0, 19 September 1996. http://vizier.u-strasbg.fr/doc/asu.html

5. F. Murtagh, "Computer networking in astronomy", in M. Albrecht and D. Egret, eds., Databases and Online Data in Astronomy II, Kluwer, Dordrecht, 1995. Preprint at http://cdsarc.u-strasbg.fr/~fmurtagh/ir-course/

6. T. Koch, A. Ardö, A. Brümmer and S. Lundberg, "The building and maintenance of robot based Internet search services: A review of current indexing and data collection methods", Work Package 3 Deliverable Report, EU Telematics for Research project DESIRE. Version D3.11v0.3 (Draft version 3). http://www.ub2.lu.se/desire/radar/reports/D3.11/

7. University of Dortmund, freeWAIS-sf at: http://ls6-www.informatik.uni-dortmund.de/ir/projects/freeWAIS-sf/; WSC Group Inc., freeWAIS-sf mirror at http://www.wsc.com

8. Isite, CNIDR (Center for Networked Information Discovery and Retrieval). http://www.isite.com

9. GAIS, National Chung Cheng University, Taiwan. http://sparc2.cs.ccu.edu.tw/ for information; and ftp://leica.ccu.edu.tw/pub/ccu/gais for software.

10. B.R. Schatz, "Information retrieval in digital libraries: bringing search to the net", Science 275, 327–334, 1997.

11. University of Maryland Information Filtering Project, http://www.ee.umd.edu/medlab/filter/filter_project.html

12. D. Harman, "Overview of the third Text Retrieval Conference (TREC-3)", National Institute of Standards and Technology, Gaithersburg, MD, 1994.

13. F. Kelledy, "Query space reduction in information retrieval", PhD Thesis, Dublin City University, 1997.

14. P.M.E. De Bra, "Finding information on the Web", 16 pp. http://www.win.tue.nl/~debra/cwi-qw/article.html

15. "The Star*s Family of Astronomy and Related Resources – The StarPages", http://cdsweb.u-strasbg.fr/starpages.html
    "AstroWeb – Astronomy/Astrophyics on the Internet", http://fits.cv.nrao.edu/www/astroweb_db.html.
    "European Physics Departments Broker", http://www.physik.uni-oldenburg.de/Harvest/brokers/EurophysNet/PhysDoc/

16. G. Newby, "Visualizing information space: spatial interfaces to textual information". T. Kohonen, "Self-organizing maps of document collections". Plenary lectures at the 21st Annual Conference of the Gesellschaft für Klassifikation, University of Potsdam, 12–14 March 1997. (Proceedings to be published by Springer-Verlag, I. Balderjahn et al., eds.)

17. Ph. Poinçot, "La recherche d'information bibliographique à l'aide des cartes auto-organisatrices de Kohonen", Proc. INFORSID Conference, Toulouse, June 1997.

18. G. Bell and J. Gemmell, "On-ramp prospects for the information superhighway dream", Communications of the ACM 39, 55–61, 1996.

19. "European research network infrastructure and intercontinental connections January 1997", available at http://www.dante.net/europanet-intro.html

20. F. Murtagh, M. Louys and J.-L. Starck, "Long-term image data storage in astronomy and the issue of noise", 1997 (submitted).

21. B. Zenel and D. Duchamp, "Intelligent communication filtering for limited band-

width environments", in Proc. 1997 USENIX Annual Technical Conf., USENIX, Anaheim CA, January 1997, available at
http://www.mcl.cs.columbia.edu/publication.html
22. G. Salton, J. Allan, C. Buckley and A. Singhal, "Automatic analysis, theme generation, and summarization of machine-readable texts", Science **264**, 1421–1426, 1994.
23. SIMBAD, http://cdsweb.u-strasbg.fr/Simbad.html
24. J. Callan, "SIGIR-96 workshop on networked information retrieval – conference report", in RIR – Rundbrief Information Retrieval, No. 5, 10–12, Dec. 1996.
25. P. Francis, "Ingrid: a self-configuring information grid",
http://www.ntt.co.jp/ntt/soft-labs/ingrid/overview.html

ELECTRONIC PUBLISHING:

THE ROLE OF A LARGE SCIENTIFIC SOCIETY

B. BEDERSON
Department of Physics
New York University
4 Washington Place
New York NY 10003, USA
bederson@acf2.nyu.edu

AND

H. LUSTIG
American Physical Society
304 Chula Vista Street
Santa Fe NM 87501, USA[†]
lustig@aps.org

1. INTRODUCTION

The American Physical Society (APS), with over 40,000 members, of whom about 7,000 reside outside the United States, is the largest society of physicists in the world. At the same time it is a major publisher of the world's physics literature. The seven sections of the *Physical Review, Physical Review Letters* and *Reviews of Modern Physics* now make up about 100,000 print pages per year. Well over half the authors of the close to 14,000 articles published each year work outside the US. Because of the growth of physics research in Europe and in other parts of the world and the perceived preeminence of the Physical Review journals, submissions from abroad have been increasing at an astounding if not alarming rate.

The APS journals now serve thousands of scientists who are not members. Conversely, the APS has thousands of members who do not publish in its journals, or for that matter anywhere. Members who don't publish in the APS journals as well as those who do expect their Society to serve them

[†]Home address.

Astrophysics and Space Science **247**: 41–53, 1997.

also in other ways than publishing. The intensely democratic structure of the APS and its variegated mission and membership provide opportunities as well as constraints for its journals. They must be and are responsive to authors, readers, referees, editors, librarians, university and laboratory administrators, multifarious subject matter communities and cultures, and the Society's governors and officers.

Even before the advent of electronic publishing, the APS journals faced many challenges. These include:

a) Covering all the important areas of physics – both traditional and emerging.
b) In each field, publishing a leading journal.
c) In an era of exponential growth in the literature, helping readers by vetting the papers that are submitted not only for their correctness but also for their importance.
d) But, in deciding which of the submitted papers to publish, being fair and being seen to be fair to all aspiring authors.
e) Helping authors to improve their papers, before publication, both in substance and style.
f) Conducting the refereeing, production and distribution process with the greatest deliberate speed.
g) Maintaining the journals in an increasingly difficult economic climate.

It is evident that while our journals have done at least as well as those of most other publishers in most of these areas, and a great deal better in some, we have not been completely successful in meeting all the expectations of all members of our community and indeed all of our own expectations.

The advent of electronic publishing has added several new challenges and expectations.

a) The almost limitless technical possibilities of cross-referencing, of hypertext, of display of data and of selective use of data and information.
b) The perception and reality that research results can be disseminated more quickly electronically than on paper.
c) The perception, and, arguably, reality that an electronic journal and, a fortiori, information not organized into journals, can be disseminated more cheaply, as demonstrated through self-publishing and e-print ventures.
d) The problem of archiving.
e) The complications of maintaining copyright protection for authors and publishers.
f) The exacerbated economics of publishing refereed and edited journals or articles electronically.

2. THE APS INITIATIVES IN ELECTRONIC PUBLISHING

2.1. SUMMARY

In the broadest meaning of the term APS has been involved in electronic publishing for well over twenty years, if one includes other aspects of computational techniques besides on-line distribution. The most important of these activities are the production and maintenance of very comprehensive referee and author data bases and the longstanding and continuing efforts in support of RevTex. Outside the area of journal publishing APS is making extensive use of electronic technology in its membership, meetings, and science and society programs. These activities will not be described in this article.

Beginning in the late eighties, APS empaneled several task forces of expert volunteers to study the status and future of electronic information systems and to recommend what the Society should do to put its publications on line. The most technical and complete of these studies, the so-called Loken Report, produced a vision of a fully electronic future by the year 2020.

The realization that our journals needed to go on line in a much shorter time span began to emerge in 1992. In 1994 we set a goal of having all of our journals on line within five years. External and internal pressures and developments steadily worked to foreshorten this deadline and we now expect to meet it before the end of 1997. *Physical Review Letters Online* was launched on 1 July 1995 and *Physical Review C* and *D* and the *Rapid Communications* section of *B* followed in 1996.

Also in 1994 APS realized the desirability, indeed the need, to produce an electronic data base for its past as well as present and future journal articles, which will become the core system for linking with other archives. Consequently, in collaboration with the Los Alamos National Laboratory, we embarked on the creation of *Physical Review On Line Archives* (PROLA). Having taken note of the success of the Los Alamos XXX preprint server initiated by Paul Ginsparg, APS in October of 1996 launched its own server. The principal motivation for establishing an APS server was to offer authors a convenient electronic means to submit to the APS journals, or simply to post papers in any area of physics as a service the world-wide physics community. Articles posted and submitted in this manner will be readily accessible to referees and editors during the review process. This facilitation is part of a plan to make the entire editorial operation electronic, to obviate the need for a paper trail, and to reduce delays and costs.

These four initiatives will be described in more detail in the sections that follow. Other projects that APS has undertaken, but which space limitations prevent us from describing further, include the adoption of an overall publishing strategy based on the Standard Generalized Markup Language (SGML) and the provision of a set of Web-based tools that will enable authors to create papers electronically with the automatic generation of the SGML structure; collaboration with the Naval Research Laboratory in its TORPEDO project(*The Optical Retrieval Project: Electronic Documents Online*) to make on-line versions of APS journals available to its researchers via an internal net called Infoweb; and collaboration with the American Institute of Physics (AIP), the (U.K.) Institute of Physics, Elsevier Science, and Chapman and Hall in the CoDAS Web, an electronic Condensed Matter Alert System covering some sixty journals in that field.

2.2. PHYSICAL REVIEW LETTERS ONLINE (MARIA LEBRON)

Physical Review Letters' policy of publishing only the most timely and significant research, as articles not to exceed four printed pages, a rejection rate of over 60%, the highest citation rate of any physics journal devoted to original research, and its more than 5000 subscriptions have made it, unarguably, the most prestigious, important, and used physics journal in the world. It was therefore natural to make bringing it on line the highest priority. The road to doing so has been anything but smooth and provides something of a cautionary tale for the APS and perhaps for other publishers.

In 1993 and 1994 the APS evaluated its own capacity as well as that of some two dozen putative service providers to undertake the on-line preparation and distribution of *Physical Review Letters*. In December of 1994 we selected the Online Computer Library Center (OCLC) for the job. This non-profit corporation was already distributing several science journals on line and had contracted with AIP for the on-line publication of *Applied Physics Letters*.

After intensive preparation *Physical Review Letters Online* was launched on 1 July 1995 with 236 individual (member) and 323 institutional (library) subscribers. It used the OCLC software package called GUIDON, a PC-based, proprietary, graphical user interface that could display, on screen, tables, text, simple mathematics, and images with typescript accuracy. The Internet was the delivery mechanism. In mid-August of 1995 a new deliverable that could make use of the World Wide Web was released, and in late September GUIDON was extended to the Macintosh platform.

From the beginning there was a problem of dealing with the high volume of mathematics-intense material that appears in the journal. In addition to improving the technology for doing so, a methodology was developed for releasing an issue with the articles that were ready and noting those that could not be loaded at the time due to software conversion problems.

Another vexing issue was the inability of the OCLC deliverable to provide a true Table of Contents.

Initial reaction to the on-line journal was mixed. Some users were very pleased with the functionalities of GUIDON and some were not. Use of the journal via GUIDON climbed from 6842 minutes of connect time in July to 16,623 minutes in December 1995. Total GUIDON connect time for the first six months of 1995 totaled 98,684 minutes. However, the preference for the Web among sections of the physics community and the initial restriction to the PC platform provoked criticism of the enterprise.

The launching of the Web version brought with it an expansion of the audience that was reached, but unfortunately also of the problems with the on-line display of mathematics. This Web deliverable was initially based on a conversion of the entire file into HyperText Markup Language (HTML). In this scheme mathematics had to be displayed as images. This created a secondary set of conversion problems, the most common one being the absence of an image to represent a specific character. As the image library grew this became less troublesome. However the problem of slow transmission due to the high number of image files increased.

Still, the release of the Web version resulted in a surge of activity. The first two weeks of the Web deliverable recorded 12,586 minutes of connect time. By 31 December 1995 total Web connect time was 87,431 minutes. At that time *Physical Review Letters Online* had 795 individual and 440 institutional subscribers.

As a result of the realization that the mathematics problem was not going to go away soon, a simplified Web deliverable was developed. This calls for the conversion into HTML of the article's bibliographic record and references only. The complete article is then delivered via Adobe Acrobat reader using a portable document format (PDF). The simplified Web deliverable was released in February 1996. A problem arose with the use of PDF files: PDF viewers are not available for some of the flavors of UNIX that are in use, particularly in the high-energy physics community.

By the end of May 1996, *Physical Review Letters Online* was well into a production mode with 100 percent article loads being achieved for many issues. However, in what constituted something of a bomb shell, OCLC announced to its clients that GUIDON was being discontinued effective 1 January 1997 and that the new Web-only deliverable would become part of a larger multi-journal project. This presented APS, as well as AIP, with the need to make new plans for the delivery of their letters journals. AIP had already begun to strengthen its on-line publishing staff and operations and had initiated discussions with APS for the on-line publications of sections of the *Physical Review*. It was therefore natural to turn over the distribution of Physical Review Letters Online to AIP. The new service began on 1 January 1997. The technology, based on PDF files, is the same as the one that will be described in the next section for the Physical Review journals. Although it initially lacks some of the features of GUIDON, we expect that the AIP deliverable will meet the needs of the physics community better in the long run.

There have been several lessons from the first eighteen months of *Physical Review Letters Online.* The physics community is highly computer literate, inhomogeneous, but generally demanding. From the individual users we learned that the ability to print the electronic version with the same look-and-feel as the hard-copy version is important, and that the system needs to be simple yet robust enough to meet the needs of a very diverse readership. Furthermore users want one-stop-shopping, with the ability to search across many volumes of a journal and many interlinked journals. These users are not tolerant of slow transmission and they prefer browsing the World Wide Web with Netscape.

From the institutional subscribers we learned that user ID and password access control do not work well in some environments and that Internet Protocol (IP) access is often preferred. Furthermore, although the financial constraints of some institutions did not allow them to pay the $250 required to continue access for 1996, a number of libraries are not ready to utilize on-line journals even if access is provided for free.

Physical Review Letters Online has been an important learning experience for APS and for the physics community. We expect to put the technological, sociological, and economic lessons that were learned to good use as we expand our on-line offerings.

2.3. THE PHYSICAL REVIEW JOURNALS AND REVIEWS OF MODERN PHYSICS (ROBERT KELLY)

At the end of 1996 *Physical Review C - Nuclear Physics, Physical Review D - Particles and Fields* and the *Rapid Communications* section of *Physical Review B - Condensed Matter Physics* were available on line. All are based on HTML, PDF, and (eventually PostScript, and are distributed over the World Wide Web. Unrestricted, free access is provided to the table of contents of each issue and to an HTML page giving the title, authors, and abstract of each article. The on-line offering also provides an advance listing of articles to be published. Where we have an author-supplied abstract in electronic form, the abstract is posted with the advance listing. Access by subscription applies to the published articles themselves, to PDF and Post-Script and to an HTML file with links to references and e-print archives.

All APS electronic journals can be accessed through the APS home page, at `http://www.aps.org`, under the Research Journals button.

Since the *Rapid Communications* section of *Physical Review B* was already available to members as a separate print journal, it was a natural candidate for the creation of an on line version. It was launched in March 1996 and it is available free of charge to members who subscribe to either part of *Physical Review B*. The articles are in PDF format. For browsing there is a complete set of titles, authors, abstracts, and references in HTML format. All items are linked from the tables of contents, which can be found at `http://publish.aps.org/PRBO/prbohome.html`. An advance listing of all papers to be published in *Physical Review B* is available, up to three months before, at `http://publish.aps.org/DLO/DL_LIST_B.html`. Complete listings of the tables of content from volume 52 (July 1995) can be found at `http://publish.aps.org/PRTOC/hometoc.html#prb`.

Physical Review C Online, which can be found at the URL `http://publish.aps.org/prcintro.html`, was launched on 1 July 1996, with articles dating back to January 1996 (Volume 53, No. 1). It is offered through the World Wide Web and includes the following features: access to the information before the print issue is mailed; a PDF file of each article; browseable current, past, and future tables of contents; searchable current and previous bibliographic records (title, authors, abstracts and PACS numbers); and the ability to print articles with the look and feel of the hard-copy version. *Physical Review C Online* allows subscribers to browse HTML versions which contain links from the individual titles to the abstract of each title. From the abstracts subscribers can browse the PDF version of the articles, including the figures, tables, and references. Subscribers can also browse the PDF versions of the tables of contents which contain links to the PDF versions of the articles as well. Access and subscription information is available at `http://publish.aps.org/prcintro.html`.

The design and timing of *Physical Review D Online* is a particularly vivid example of APS' dependence on and responsiveness to its community: in this case the very computer literate and vocal high-energy physics community. A strong stimulus came from the modus operandi and success of Paul Ginsparg's e-print server at Los Alamos. A number of technological approaches were considered. *Physical Review D Online* is now available at `http://publish.aps.org/PRDO/prdohome.html`. It offers the following: access to *Physical Review D* several weeks before distribution of the printed journal; a file of current and previous tables of contents; a file of each article's abstract that includes the title, list of authors, PACS numbers, and references with links to SLAC, SPIRES at the Los Alamos National Laboratory e-print archives; a PDF file of each article; an advance listing of papers to be published in future issues; and the ability to print a replica of the hard-copy version. Readers can browse the tables of contents which contain links to the abstracts and references of each article and the PDF versions of the entire articles. The articles are currently available only in the PDF format. However we expect to be able to offer PostScript files to our users in the future.

Physical Review A - Atomic, Molecular, and *Optical Physics, Physical Review B - Condensed Matter Physics,* and *Physical Review E - Statistical Physics, Plasmas, Fluids* and *Related Topics,* will join the existing electronic journals during 1997. They will be initially distributed in the PDF format and, like those already on line, will be processed for APS by AIP. *Reviews Of Modern Physics* is also expected to go line in 1997; the format and the delivery service have not yet been decided on.

2.4. PROLA – PHYSICAL REVIEW ONLINE ARCHIVES (TIMOTHY THOMAS AND ROBERT KELLY)

Since 1893 the *Physical Review* has published approximately 1,700,000 pages of carefully reviewed, edited, formatted, and printed material – an unmatched repository of physics knowledge. APS in conjunction with Los Alamos National Laboratory (LANL) is developing the Physical Review Online Archive (PROLA). It will be an easy-to-use, fully searchable and navigable desktop system that provides on-screen-viewable and printable versions of all *Physical Review* articles from 1893 to the present. The system is based on the World Wide Web and the Wide Area Information System (WAIS) and the collection will be available over the Internet world-wide.

A prototype is already in use at LANL. It offers full functionality for the years 1989 to 1993 for all sections of the *Physical Review,* for a total of 51,000 articles. An additional 60,000 articles covering the years 1985-1988, 1994, and *Physical Review Letters* for 1985-1994 are available in abstract and searchable text versions. APS is now working with the Naval Research Laboratory to obtain the scanned images of these articles. They will be linked to the existing text versions, giving full coverage for 1985-1994, with 1995 soon to follow.

PROLA is centered on the delivery of scanned images of individual pages. This provides the accuracy and speed required of an archival system. The technological problem is how to down-sample a 300 dots per inch (dpi) bit-mapped page image to a 100 dpi anti-aliased byte image that is easily readable on a typical monitor, while at the same time offering the original, high quality dpi image suitably packaged and compressed for printing. PROLA accomplishes this by preprocessing all images to produce the desired versions and then hyperlinking to these versions from a single document information page.

The scanned images are not searchable by existing technology, so PROLA also needs a text version of each article. Since 1985 APS has retained troff, RevTeX and SGML versions at different levels of completeness. These versions, however, cannot be used to produce an accurate representation of the published article, since they often do not have the figures and tables, nor are the correct macros and style forms still available. In any case the published version was often produced by cutting and pasting from different files in unknown ways. For these reasons the text versions can in no way be considered sufficiently accurate for an archive. Nevertheless, PROLA offers these versions in conjunction with the image, to allow searching the full text and permit access for non-graphical Web users.

The necessity of providing the large page images makes for considerable storage and speed of delivery problems. PROLA solves them by utilizing the tape/disk storage system at Los Alamos in conjunction with a large local cache on the Web server. Whenever a document information page is requested by a user, the full set of page images is immediately retrieved from the tape system and moved to a local cache. A user who wants to read the article on line is then prompted with a set of small thumbnail images and asked to select a page for viewing. By the time thumbnail image has been selected, the page images have been moved and can be delivered promptly. A similar system works for delivering the printable images.

PROLA offers important enhancements unavailable with paper archives. The ability to do complex, boolean full text searches is clearly the most critical, but the ability to hyperlink to articles from the references may be equally important. PROLA posts two lists on each document information page: the first provides a link to all PROLA articles that are referenced in the current article; the second links to the later papers that reference the current paper, including any errata that may have appeared. While PROLA at present offers hyperlinks only to articles in the PROLA system, a system that locates and lists all references to archived articles from the 19986 on-line APS journals is under construction. Furthermore, planning for offering links to references in non-APS on-line journals has begun. This service will depend on the willingness of the other journal to link reciprocally to PROLA.

PROLA also offers "feedback searches", i.e. the ability to locate articles that share similar rare words with the current article and are therefore likely to deal with the same topic. This lets the reader be assured that all relevant *Physical Review* articles have been located. PROLA also keeps track of each time that a document information page is accessed and gives that count when a search is completed. This lets the user quickly identify the most popular articles, which may also be the most useful. Another feature is a direct link to a PACS subject dictionary. We hope that this service cab be enhanced to add a subject browser which will augment the existing table of contents browser.

From a technical point of view PROLA is nearly ready for a global trial and for feedback from the community of users. We have not yet addressed the problem how to find the needed financial support for it. But there is little doubt that PROLA has the potential for becoming one of the most important and beneficial electronic offerings of the APS.

2.5. THE APS E-PRINT SERVER (ARTHUR SMITH AND MARK DOYLE)

Preprints have been an important method of disseminating research results in physics for many decades. These early, unrefereed and often as yet unsubmitted papers became a primary means of rapidly exchanging information, particularly in the field of high-energy theory. The development of TEX in the mid-80's allowed for high quality typesetting in a virtually platform-independent environment and by the late 80's e-mailed TEX was used to speed up the distribution of preprints.

In 1991 Paul Ginsparg of LANL created an automated clearinghouse for preprints in high-energy physics. Authors could submit their TEX files by e-mail to the "e-print server" and subscribers would receive a daily mailing of the abstracts of the papers that had been submitted the previous day. The papers could then be retrieved by interested readers via e-mail. The original idea was to maintain preprints for only three months. But it was soon decided to maintain the papers indefinitely, if only to obviate the need for having to accumulate stacks of preprints on desktops and in filing cabinets. Thus was born the Los Alamos E-Print Archive. Subsequently archives were added in many other fields of physics and mathematics, but high-energy theorists remain the predominant users of "Los Alamos XXX".

After two years of study by several taskforces and committees the APS decided to mount its own e-print server. This decision was based on the realization that, while the preprint culture was not pervasive in all fields of physics, nevertheless this model of communication could offer benefits for the whole community which the APS exists to serve. Another motivation was to enhance the modalities of submission and refereeing and production for our journals.

E-prints are readily available and immediately accessible to anyone with an Internet connections. Readers can use computer-based methods to search bibliographic information. E-prints in original text formats can also be enhanced by allowing conversions to reader-specified formats, and even the automatic propagation of external and internal linking information into a hyper-linked, display-oriented format.

Another goal of the APS e-print server is to investigate methods for making Word, WordPerfect, and other word-processor based formats usable for on-line distribution. This may require conversion of these formats into Postscript, PDF, or other document display formats. One eventual solution may lie in SGML that is now being used for the production of APS journals and which provides a source language for very flexible conversions on output.

The Web provides a reliable means for the transfer of files, particularly binary files. Thus the e-print system could become the primary mechanism for submitting papers to the APS journals for publication (Submission from the Los Alamos and other public We-based sources will also be encouraged). Papers can then be transferred electronically to and from referees. In addition the e-print system will allow posting of materials that cannot be translated into print.

From July through September 1996, the APS e-print system was publicly available in prototype form The server went "officially" on line in October 1996 and included enhanced functionalities, such as full-text searching of the articles and automated conversion of TeX-based articles to PostScript and PDF.

Present plans are to remove articles from public access on the APS e-print server after two years. They reflect the belief that preprints are usually valuable only for a brief period and that it is not desirable to have unrefereed papers publicly available in perpetuity. They also reflect a concern about the relationship of these preprints to articles that have been refereed and published in the journals. With the journals available electronically, the e-print version may be replaced simply by a pointer to the journal version. Some members of the community maintain that a finite life-time for e-prints neglects those that are never intended for publication in a refereed journal, such as graduate student theses, technical reports, and conference proceedings. A separate "permanent" e-print area may be made available on the APS server for such materials.

Unlike the Los Alamos server, which consists of a large number of separate "archives", the APS server is intended to remain as a single database. Nevertheless there is a legitimate concern of duplication with XXX, or of having to look in multiple places for the latest e-prints. We are investigating ways of integrating our search and notification system with the Los Alamos one, possibly through cross-posting between the two. The APS e-print server may be accessed at http://www.aps.org/eprint. Comments are also welcome by e-mail at eprint-adm@aps.org.

3. THE FUTURE OF TRADITIONAL JOURNALS IN THE ELECTRONIC AGE

In the culture of the e-print server and the "free internet", what is the future of traditional scientific journals? How can they compete with the quick transmission of information afforded by these new approaches and technologies?

. It must be recognized that it is not only the technology of editorial mechanics, printing and distribution by mail that is responsible for the delay in traditional scientific publishing and the speedier and less costly availability of e-prints. Rather it is also the complex and time consuming process of selecting and engaging referees, the many cycles of interaction between referees, editors and authors, and the editing of the articles. Even after having, through successful reengineering, made the process entirely paperless, a carefully and fairly refereed and well-edited journal or article will take longer to reach the readers than unselected, unvetted, unedited and unimproved e-prints.

Therefore scientific societies, such as the APS and other traditional publishers, in addition to making their print journals available electronically, face the issue of shortening the non-technological aspects of publishing, by speeding up the refereeing process, allowing less opportunity for rejoinders and appeals by authors and curtailing or eliminating stylistic and other editing.

In fact, some hold that not only editing but even refereeing should be given up. Traditional publishers, such as the APS, some say, must adopt the e-print approach or go out of business. This imperative, we believe, is not shared by most providers and consumers of scientific information. The value of vetting and certifying the fraction of information from the enormous pool being produced, that is worthy of attention, and of making it more readable, is simply too great to allow giving up on the refereed journal or article just yet.

Indeed this is recognized by many of those who thrive in the e-print culture. A group of American high-energy theorists has been considering refereeing papers directly from the XXX e-print server and simply endowing the fraction that is accepted with the seal of approval that publication in a refereed journal normally entails. There would be little or no editing.

These retrenchments from traditional publishing practices – reducing the editing and not distributing anything – would result in somewhat lower costs, but not by much. For the traditional ways of doing business, some of the costs are independent of whether the mode is print or electronic, others are reduced by switching to electronic distribution, while others, having to do with the ease of uncontrolled access, are exacerbated. The economic challenges that the APS and other publishers are faced in this era of incipient electronic publishing are examined in another chapter of this volume.

One of the steps to meet the very real economic challenge , whose motivation and implications transcend economics, is APS' decision to reduce the acceptance rate for submissions to the *Physical Review* below its current already relatively low numbers. While, even in the face of increasing submissions, this will permit APS to hold down the price increases for its journals, we also believe that being more selective will increase the quality and usefulness of the journals, albeit at the expense of the peace of mind of embattled editors and disappointed authors. The free and easy availability, via e-print servers, of information, important and unimportant, correct and erroneous, makes it acceptable and desirable for refereed journals – whether they are electronic or print – to be more selective.

Disparate views about the future of scientific publishing have sometimes been expressed by members of the three groups most intimately involved in and affected by it: scientists, publishers and librarians. The perception of the roles of these three players and the roles themselves are changing, fueled in part, by the advent of on-line distribution of scientific information. No one has as yet proposed that the scientists should be eliminated (although their ranks may well be thinned by the present reduction in research funding by governments, universities and industry). In contrast, either and sometimes both of the other two groups have been nominated as candidates for extinction. It is possible that the whole job of creating, vetting and disseminating scientific information can be done by the working scientists, but only if they assume the publishing function and burdens of publishers and the role of providing access exercised by librarians. Short of this, with some changes in policies and procedures, the three groups can cooperatively assure the survival and enhancement of the dissemination of the results of research in an era of technological and economic change. Nonprofit scientific societies, with a publishing mission, such as the American Physical Society, represent and answer to all three groups. They are in a favored and even in a mandated position to lead the effort.

Notes

Much of Section 2 of this paper is based on or was taken, in an edited version or verbatim, from articles in a special supplement "APS Online" to the November 1996 issue of *APS News*, the American Physical Society's monthly newsletter. The names of the authors of the articles are given after the titles of each initiative.

The views expressed are those of the authors and not necessarily of the American Physical Society.

THE AAS PROGRAM OF ELECTRONIC PUBLICATION

P.B. BOYCE
American Astronomical Society
2000 Florida Avenue, Suite 400
Washington DC 20009, USA
pboyce@aas.org
http://www.aas.org/~pboyce

1. Introduction

At the outset we need to make clear what we mean by electronic publishing. Heck (1996a, 1996b) has been saying for some time that "publishing" means much more than making the traditional journals available electronically. We agree. Electronic publishing must include everything we distribute over the net; everything from peer-reviewed journals, to basic data, to general information. At the AAS, we have seen the effects of this new world in which it is possible to bring a whole continuum of digital information to the desktops of working scientists and scholars. Full electronic access to a wide range of information is already producing a new working environment. I predict that our newly developed ability to access not only the peer-reviewed journals, but also original data, intermediate results, tables of information and unrefereed preprints will change the way research is done in subtle, but powerful ways.

The ability to communicate and work effectively with colleagues thousands of miles away is already changing research partnerships. The ability to work with like minded colleagues wherever they are shatters the traditional fetters of geographical proximity and brings a fresh attitude, new stimulation and remarkable creativity to astronomical research. I believe this trend will grow, and may, in the end, be the most important result of the electronic revolution.

But this is a topic to be explored in a different paper. Suffice it to say that it is the whole range of digital information which must be considered as we proceed into the electronic future. For this paper, we will limit our comments to the AAS programs in the transfer of scientific information,

Astrophysics and Space Science **247**: 55–62, 1997.

leaving out electronic publication of items such as career opportunities in the AAS Job Register, meeting information, Society organization, staff lists etc.

2. Starting with the Meeting Abstracts

As the first step into electronic publishing (Boyce *et al.* 1992), the AAS began the electronic submission of abstracts and papers for meetings of the AAS in 1992 and was immediately embraced by the community. Electronic submission is now the normal procedure. Meeting programs and abstracts are routinely made available before the meeting in electronic form. Four years ago, the percentage of electronically submitted papers for the American Astronomical Society had climbed above 99%, and it is now about 99.8% (2 paper abstracts per thousand submissions). This is a remarkable level of acceptance by the community. I believe the community will embrace the electronic peer-reviewed journals to nearly the same degree within a year or two after the full literature becomes available electronically. The only caveat to this statement is that the electronic journals must be substantively superior to their paper counterparts for this to happen.

Electronic meeting abstracts were simply the first step taken by the AAS toward bringing the peer-reviewed journals on line, but the experience gained in developing the "publishing" process needed to accept, process and display the abstracts proved invaluable as we moved to the harder task of bringing out an electronic version of our journals. The planning and development phases of the AAS Electronic Publishing project have been described elsewhere (Boyce & Dalterio 1995, Dalterio *et al.* 1995, Boyce & Dalterio 1996, Boyce 1996 and Boyce *et al.* 1996).

The major lessons we learned in this early stage can be summarized by four statements.

First, dealing with electronic documents by hand is usually very expensive in time and requires relatively expensive expertise. Therefore every effort needs to be made to handle documents automatically.

Second, the whole process is much more interdependent than is the case for handling paper documents. Every step of the process leading from electronic submission to the preparation of the electronic product depends upon all the preceding steps.

Third, the standard of precision required to be able to handle electronic documents automatically is much higher than if documents are being handled by humans. However, automatic checking can be used to catch author mistakes and improve the accuracy.

Fourth, as a consequence of the first three statements, the whole process, from end to end, must be planned around the electronic environment, tools and products. One can almost never successfully convert just part of the publishing process into electronic form.

Evan Owens (Walters 1997) of the University of Chicago Press put it correctly when he said at a recent Seybold conference, "We've been spending the past several years, not so much on online journals as on building the infrastructure for the future."

The importance of this approach will become apparent in the next year or two as the electronic collection grows and as the tools, technology and products evolve. Electronic materials which have been developed with a robust underlying architecture, using early conversion to SGML, insistence on the use of name resolution techniques for links, avoiding proprietary software systems, and including the use of bibliographic databases will age and change gracefully. Journals and materials which have not been built upon a robust underlying infrastructure will not migrate easily to new technologies and standards, and will no longer be accessible. Some materials on the WWW, such as meeting information, have intrinsically short lifetimes. For them, longevity is not an important consideration. But most scholarly materials, by their very nature, are expected to be available into the indefinite future. For them, a robust understructure is imperative.

3. The Peer-Reviewed Journals

The AAS has been producing successfully an electronic version the Letters section (2,500 pages annually) of the Astrophysical Journal since September 1995. By starting with the smallest "entity" that we publish, a journal which could be proofread and corrected by hand, we had the freedom to experiment and to refine the electronic production process to a degree which would have been impossible for any of our larger journals. During this process we have come to understand the great difference between electronically delivering images of paper pages and building a journal specifically tailored for electronic presentation and structured to take maximum advantage of the electronic environment. In fact, we no longer regard so-called electronic journals which only deliver page images (without links) as "true" electronic journals. Such journals are nothing more than refined document delivery systems.

The AAS has spent a significant effort on tailoring the electronic version of its journals to be useful to the reader. To date, the author is unaware of any other journal on the Internet which has as rich a set of features and links. The features of the ApJ which help to set it apart from other electronic journals are:

– Full internal hypertext linking for

- References (to reference page and back to text)
- Figures (thumbnails, captions, links to full figures and links back to the text)
- Tables (in both HTML and ASCII formats)
- External Hypertext linking to
  - Abstracts of referenced articles available in the Astrophysics Data System (ADS)
  - Through the ADS to page images of the older articles (20 years of articles available)
  - Citations to article from other electronic journals
  - Electronic articles in AJ, ApJ and other major journals when they come on line
- All links done through name resolver, not URLs (for robust links, mirror sites).
- Both HTML (for screen reading) and PDF (for local printing) formats
- SGML archive with proven completeness and reliability; translatable to new standards
- User-friendly design of the electronic version, minimization of transmission time, avoidance of superfluous graphics, careful attention to needs of users.

In addition the AAS and its electronic publisher have prepared for the incorporation of additional features and enhancements. The structure of the journal preparation process, which focuses on the preparation of a richly tagged file in SGML, even allows for the retroactive inclusion of new features. During the first two years of operation, the pace of development has been so rapid that we have reconstituted the entire set of publicly available issues about every six months.

We now provide features which were not available when we started the electronic version in September 1995. Some new features, such as links to the papers which cite the one you are reading, are major improvements. Other important improvements, such as the use of the Netscape (and Microsoft Explorer) table formats which provide more readable tables, were added to take advantage of new browser capabilities. But there are many little improvements, which may have escaped the casual reader. Among these we can list the incorporation of the expanded Netscape character set – which reduces the number of special characters which have to be represented by GIF images. All of these features have been incorporated into the complete body of available issues by automatic reconstitution of the HTML version of the electronic journal from the underlying, archival

SGML database. The capability to produce a new version through the use of translation scripts will be vital for the inexpensive maintenance of our electronic archive in the future.

We are still breaking new ground in our electronic publishing efforts. The AAS has been fortunate to be able to develop a cooperative relationship with the University of Chicago Press who have been willing to respond to feedback from users as well as new AAS requirements, while at the same time continuing the daily production process of putting out several major research journals. Only in this way have we been able to incorporate new developments and continue to expand the features of our electronic journal.

With eighteen months of experience in publishing the Letters, the AAS has just brought the complete ApJ (25,000 pages annually) on line in January 1997, and will institute access control in April 1997, allowing access only by licensed subscribers. With our major journal completely on line (joining the ApJ Letters, New Astronomy, and the A&A Supplement Series) the astronomical community now has over 40% of the new peer-reviewed literature available electronically.

4. Links, and the Importance Thereof

Upon making the ApJ Letters available electronically we immediately confirmed through reader feedback that one of the most important features of an electronic journal is the ability to link immediately and easily to the articles referenced by the author. This feedback has been so strong that we now believe that the system of references and citations connected to each electronic article is one of the most vital attributes of an electronic journal, far exceeding the importance we originally gave to this feature. Robert Kelly (1996) of the American Physical Society has stated that the links may be even more important than the text, the general contents of which may already be known to the reader through the medium of electronic preprints.

While it is relatively easy for an electronic journal to include links to other articles in the same journal, it is more difficult to link to articles in other journals. We have developed a solution which uses the bibliographic information and abstracts of the astronomical literature, often called the metadata by the electronic publishing community, which is available in searchable form at the Astrophysics Data System (ADS). In addition to being a valuable resource in its own right, the ADS is the central mechanism by which the electronic literature and data in astronomy are linked together. We use a standard system of naming, which was developed a decade ago by the electronic data centers in astronomy, now called the "bibcode." Using a standard name it becomes easy to produce the links among different scholarly papers, different journals, the information on different astronomical objects and the actual observational and calculated data upon which our knowledge of the universe is based.

In addition to having the new astronomical literature available in electronic form, much of the last twenty years of the astronomical literature in astronomy is also available electronically through the ADS in the form of bitmaps. Of course the new literature links to both the abstracts and the full text presentation of the historical literature. But even more important, the ADS in cooperation with the AAS, is now providing the capability to link between papers in the historical collection of papers in page image format, a capability not usually associated with page image formats. But now readers of a historical paper have the capability to call up the reference lists in HTML format so they can now follow the links to the abstracts of the referenced papers and their associated page images of the full text). They can also get a list of the papers which cite the subject article and can thus link to them as well. The last twenty years of the scholarly astronomical literature is part of this fully interlinked system and the AAS electronic journals integrate smoothly with this resource as well.

5. The Urania Collaboration

For years several major data centers in astronomy have been providing citations from their data tables to the published literature from which the data were extracted. Once the peer-reviewed journals began to come on line, it was natural that they should link directly to the electronic literature. As a consequence, the publishers, the data centers and the metadata provider (the ADS) refined slightly the standard bibcode naming convention and adopted it for naming the articles now available electronically. The bibcode has the advantage of being calculable automatically from the standard reference form for journal articles. Scripts have been developed to do this and then to query the ADS to check the validity of the reference. If a link appears in the ApJ, it will work. As mentioned above, the ApJ (and also the ADS) links are based upon the use of Universal Resource Names (URNs) instead of URLs. Not only does this ensure the stability of the links, but it also makes possible the rational operation of mirror sites for the ApJ (and for the ADS) in Europe and elsewhere.

By adhering to a few conventions on naming, the use of URNs and the provision of information automatically to inquiries from each other, a small group of digital information providers in astronomy have developed a working prototype of a distributed digital library. At the suggestion of the author, we have agreed to call ourselves the Urania Collaboration (Boyce 1997). Urania, by effectively linking a number of resources, shows how much can be accomplished using simple protocols and conventions in a distributed environment. But, even more, Urania demonstrates that publisher and other information providers have much to gain by working together. The WWW (remember the third word is Web) is an inherently linked environment. Individual journals or other information providers which do not reach beyond their own boundaries will not survive in the new electronic environment.

6. The Future

We know the future will be a time of great change. The electronic tools available to us will continue to change at a rapid pace, driven inexorably by the expected economic benefits to those companies which provide tools for the use of the WWW by millions of people. If the scientific community can adapt to this pace and can make use of the mass market tools, we will have capabilities at our fingertips which have, until now, not even been dreamt of. It will be up to us to take advantage of these developments. The problem will be to ensure that our material survives and is accessible. This is a source of some concern. But, if we have built our scholarly material well enough, we will be able to adapt and migrate our products to the new technology. This should be our goal.

Acknowledgments

This work was carried out in part with support from the US *National Science Foundation* to the *American Astronomical Society*. Recent work has been accomplished in part under a Cooperative Agreement with the US *National Institute of Standards and Technology*. I am grateful fur the continuing support of my collaborators, Evan Owens from the *University of Chicago Press* and Chris Biemesderfer of *ferberts associates*.

References

1. Boyce, P.B. 1966, *Computers in Physics* **10**, 216
2. Boyce, P.B. 1997, URL: http://www.aas.org/Urania/
3. Boyce, P.B., Biemesderfer, C. & Owens, E. 1996, *Vistas in Astron.* **40**, 423
4. Boyce, P.B. & Dalterio, H. 1995, *Vistas in Astron.* **39**, 209
5. Boyce, P.B. & Dalterio, H. 1996, *Physics Today* **49-1**, 42
    – URL: http://www.aas.org/~pboyce/epubs/pt-art.htm
6. Boyce, P.B., Dalterio, H. & Biemesderfer, C. 1992, Three Year Plan for Developing Electronic Publishing at the American Astronomical Society
    – URL: http://www.aas.org/Epubs/webinfo/Plan/epplan92.html
7. Dalterio, H., Boyce, P.B., Biemesderfer, C., Warnock III, A., Owens, E. & Fullton, J. 1995, *Vistas in Astron.* **39**, 7
8. Garrett, J. & Waters, D. 1995, Preserving Digital Information. CPA/RLG Report
    – URL: http://www.rlg.org/ArchTF/tfadi.index.htm

9. Heck, A. 1996a, *Vistas in Astron.* **40**, 303
    – URL: http://cdsweb.u-strasbg.fr/~heck/itvistas.htm
10. Heck, A. 1996b, *Vistas in Astron.* **40**, 365
    – URL: http://cdsweb.u-strasbg.fr/~heck/stia.htm
11. Walters, M. 1997, Online Journals: Print Publishers Move from Pilot to Full Rollout,
    in *Seybold Report on Internet Publishing* **1-6**

THE PROJECT OF ELECTRONIC PUBLICATION
OF ASTRONOMY AND ASTROPHYSICS

H.J. HABING
Sterrewacht Leiden
Postbus 9513
NL-2300 RA Leiden, Netherlands
habing@strw.leidenuniv.nl

AND

J. LEQUEUX
Observatoire de Paris
61, avenue de l'Observatoire
F-75014 Paris, France
lequeux@mesioa.obspm.fr

1. Introduction

Like all the important primary journals in astronomy, *Astronomy & Astrophysics* (A&A) is turning to electronic publication. The Board of Directors of A&A has decided on the principle at its last meeting in May 1996, and we are implementing it progressively with the help of our publishing houses, *Les Éditions de Physique* and *Springer-Verlag*.

2. Why electronic publication?

Electronic publication is unquestionably the future for primary scientific journals. Some arguments are common to all journals: ease of consultation (when the Web is not saturated!), possibilities of getting print-quality copies, low-cost color figures, ease of personal archiving, and easy bibliographical searches. Another advantage which has not yet been much exploited is the possibility of multi-media display, including movies, sound, etc. Sooner or later, the paper version will be reduced or even perhaps disappear, with corresponding savings on paper and space on library and personal shelves.

Astrophysics and Space Science **247**: 63–68, 1997.

For astronomy, the most important aspect of electronic publication is the on-line access to bibliographical and other databases. There exists a superb tool for bibliography: the *Astronomy Abstract Service* (ADS) which contains the abstracts and even the full texts of most publications in astronomy and related fields, as well as softwares for further searches by key words or by author names. The SIMBAD database of the *Centre de Données astronomiques de Strasbourg* (CDS) contains basic data and complete bibliography for astronomical objects, as well as an extensive set of related search tools. The CDS also prepares Aladin, which will give maps with identifications around any object. Last but not least, A&A already publishes the abstracts of the papers and a large number of tables and other material purely electronically at the CDS, where they can be accessed by ftp (130.79.128.5) or via Internet (http://cdsweb.u-strasbg.fr/Abstract.html). Electronic publication should provide hyperlinks to all those facilities.

For the moment, this is only possible from an HTML (HyperText Markup Language) version, although the situation is evolving rapidly and other versions like PDF may give this service in the future. HTML also provides a faster, interactive reading of the papers, with internal links between text, figures, tables and references as well as the above-mentioned external links. Electronic journals generally provide their readers on the Web with two simpler versions: a PostScript one for printed-journal quality, a PDF one for simple, non-interactive reading. Preparation of these versions is straightforward and inexpensive, as printing of the paper issues generally uses a PostScript file; the PDF file can be produced easily from the PostScript one. Most of the journals outside astronomy do not prepare the HTML version, which is more costly mostly because the internal and external links have to be implemented. It is generally estimated that the overcost of a parallel electronic version with HTML over pure paper publication is of the order of 15%, mainly personnel costs, with some part for the necessary hardware and connections. Apparently astronomers are ready to pay for this as soon as they realize how useful is the HTML version. The journals of the *American Astronomical Society* are moving progressively to parallel electronic publication including HTML (see for the *Astrophysical Journal* http://www.aas.org/ApJ/), and a new journal, mostly electronic with a reduced paper version, has started very recently (*New Astronomy*).

3. The A&A project

We shall distribute initially all three versions of the articles mentioned above: PostScript, PDF and HTML. This has started officially for the Supplement Series as to January 1, 1997 (see http://www.ed-phys.fr). The abstracts can be read free, then if interested the subscribers (and only the subscribers) may access the full versions of the paper. The HTML version contains internal links between text, figures, tables and references, and external links to the ADS, to the electronic tables at the CDS and to SIMBAD. For the moment the SIMBAD link is implemented as a global link after the abstract, and individual objects discussed in the paper can then be accessed for simple information (free) or for complete bibliography and other informations if the customer has a subscription to SIMBAD. Later, we plan to implement links to SIMBAD and to Aladin by clicking on object names in the text. The abstracts and tables are available for free at the CDS as explained above. At present, there are only PostScript and PDF versions of the papers in the Main Journal (see http://science.springer.de/aa/aa-main.htm), but we hope that Springer-Verlag will prepare also HTML versions soon so that we can go for full electronic publication in 1998. We also plan to distribute regularly CD-ROMs containing all the material of the journal, at a rather low price.

We consider the electronic version as being an integral part of the journal. While a paper version will continue to be distributed in the foreseeable future, the electronic version contains more material: this has been the case for years for large tables which are not printed but only distributed by the CDS. It should be clear that it is the whole version including the purely electronic material which is submitted to the referee: this is especially important for the data tables which are the long-remaining part of astronomy and should be self-explanatory, with a consistent use of astronomical nomenclature. To which extent the electronic version will differ from the printed one is presently the subject of an active debate in the community. We will come back to this point later. In any case, A&A, as the other main astronomy journals, will have its quality guaranteed by the refereing system and the editors, contrary to preprint databases like the SISSA ones where everyone can put his paper freely without any quality label.

We do not want for the moment to distribute the electronic version alone. This means that only the libraries and individuals that subscribe to the printed version will have access to the electronic version. This can be managed easily: the servers of the electronic version recognise the IP number of the computer which tries to connect to it and see whether it is authorized. Access will become free after some time. As Springer-Verlag and Les Editions de Physique are not willing to maintain their database for more than two years after publication, we are preparing an agreement with the CDS which will maintain the whole database indefinitely, duplicate it to mirror sites, and transfer it to new supports according to the evolution of those supports. We also hope to be able to have mirror sites of the current issues of the journal in other countries than France and Germany, in order to ease the consultation by foreign users.

4. Financial problems and consequences

All this has a cost. We have already indicated that the overcost of parallel electronic publication with respect to the paper version is of the order of 15%, which must be supported by the subscribers as we do not have the page-charge system of some other journals, in which most of the income comes from the authors. Introducing such a system would oppose to the European habits. However the subscription prices are already high although every effort has been made to keep them at the minimum: for example, the authors must now prepare their papers in LaTeX and produce their figures in PostScript so as to minimize the costs at the publisher. In this way, the actual subscription rate per page has continuously decreased since the fundation of A&A 28 years ago. But we are reaching the limit, and while the Europeans are ready to pay for A&A since it is "their" journal, the subscription price is rightfully considered as very high by our colleagues of other countries like the USA, where they get the Astrophysical Journal at a low price (forgetting that they also pay for it with their page charges!). It is a fact of life that the number of articles we receive is continuously increasing, mostly due to the power and efficiency of observing means and of computers. It is too easy for those who are not in an editor's skin to claim that the rejection rate should be drastically increased: we are convinced that most of the material we publish is good and worth being published, and that only little savings can be made through an enormous effort from our side. But the question of conciseness arises, and it is where we can gain substantially. Is it really necessary to print *on paper* all the material which is published at present? In our opinion, the answer is clearly no, although the authors who like to see their stuff printed in a journal may think differently.

It is interesting to dig a bit more into this rather crucial subject, which is of course quite controversial. It is clear that most scientists do not read papers entirely. Usually they browse the table of contents of the journal, then if interested read the abstract, perhaps the introduction and the conclusion, and look at the figures. Complete reading is usually done only by the specialists, who may be only a dozen or so on average. Of course there are exceptionally good papers which are read entirely by many, but they are probably rare and in any case it is difficult to predict at the time of publication which papers will become classicals. Then the question arises: is it really necessary to print everything, provided that all the information exists and is preserved somewhere and is readily accessible, although perhaps slowly and less confortably than in a printed issue? Our own answer is no, although most authors may think differently as far as their own papers are concerned (not necessarily other papers!). In any case, sooner or later we will be forced to such a solution for reasons of space on the shelves, cost and ecology. It is thus time to consider that some parts of papers could be defered to purely electronic appendices, as already done for tabular material. Some papers we publish in A&A Supplements already conform to this scheme. What will be the consequences?

For the author, it will mean to conceive his article in a new way, where only the essentials will be published on paper while all the details and developments will be found in electronic appendices. Of course there cannot be a rigid rule, and there will be all intermediate between short papers whose printed and electronic versions will be identical and articles reduced to some sort of announcement of material published only electronically (a formula currently in use in the Supplements). Also, what is to be kept on paper depends on the subject: sometimes the figures are the essential part, and sometimes they can be mostly published electronically. One has to be very pragmatic and flexible in this matter.

For the user, the printed version will take less room and will be easier to consult and to browse. The problem will be the rapidity of access to the electronic version, and here we must say that the situation is rather unpredictable, so that one should be careful and make experiments. CD-ROMs will be available some time after the publication of the article and allow fast consultation of the complete material of the previous issues (including the purely electronic material).

For the librarian, it will mean also savings of space, but it is clear that the libraries will have to be equiped with terminals and CD-ROM readers. Fortunately, such equipment is unexpensive nowadays and we believe that the corresponding expenses will be more than compensated by the decrease in cost of the journal, which will also be published faster since there will be no backlog due to its reduced size.

Two categories of people may suffer from such a change: the old and the poor. The old, because they will have to change their habits. The poor (essentially the countries of the former USSR), because they will have to pay for access to Internet and for terminals and will have slow connections; on the other hand, we cannot afford anymore to serve them free subscriptions to the paper journals, as it has been done in a recent past: they will have problems in any case, unfortunately.

5. Conclusions

We shall continue to offer a printed version of A&A for the foreseable future. At the same time, there will be a parallel electronic version on the Web, which exists already for the Supplement Series. The HTML format of this version offers incomparable possibilities for on-line bibliography and searches in databases, thanks to hyperlinks to existing.and future databases.

Some material is already published only electronically, and more will probably be in the future. Quality control requires that the whole material (paper + electronic) goes through the refereing system, and perennity of the knowledge requires guarantees for indefinite preservation of the complete database. Provided this is realized, the hybrid system results in savings on paper, cost and space. Whether these advantages will balance the difficulties, cost and slowness of Internet links is not completely obvious, although we are personally convinced that they do. The enormous advantages offered by electronic publication are worth an effort by the different parties. But we are in an evolving and unpredictable situation, and we should only progress with care through a continuous dialogue between authors, editors and publishers.

THE ACM ELECTRONIC PUBLISHING PLAN

P.J. DENNING
George Mason University,
Fairfax VA 22030, USA
pjd@cne.gmu.edu
http://www.cs.gmu.edu/faculty/denning.html

AND

B.A. ROUS
Association for Computing Machinery, Inc.
1515 Broadway
New York NY 10036-9998, USA
rous@acm.org[†]

1. Introduction

Publishing has reached an historic divide. Ubiquitous networks, storage servers, printers, and document and graphics software are transforming the world from one in which only a few publishing houses print and disseminate works, to one in which any individual can print or offer for dissemination any work at low cost and in short order. This poses major challenges for publishers of scientific works and for the standard practices of scientific peer review.

The ACM aims to be one of the first scientific society publishers to cross the divide. ACM has embarked on an ambitious electronic publication plan. The plan and the reasons for adopting it are set forth below.

The ACM is the first scientific and educational society formed in the computing field (founded 1947). From the very beginning it entered scientific publishing by establishing the monthly *Communications of the ACM* and a peer review process for accepting papers into it. Over the years, its library of traditional journal-type publications has grown to the present

Astrophysics and Space Science **247**: 69–82, 1997.
© 1997 *Kluwer Academic Publishers.*

size of 17 periodicals including the one monthly, several bimonthly, and the rest quarterly. Its 79,000 members hold 55,000 subscriptions to its journals, and nonmembers hold another 13,000 subscriptions.

In the 1960s, ACM established a series of special interest groups (SIGs) that started issuing informal newsletters of their own and began to hold conferences and symposia that published proceedings. Over time, this grew into a large enterprise, featuring 90,000 memberships in 39 SIGs that sponsor 45 conferences per year and print 17,000 pages of proceedings. All told, ACM literature is growing at the rate of approximately 1 gigabyte per year. The publications of the traditional journals and SIGs constitute a large enterprise, on the order of a third of ACM's $30 million budget.

2. The Scientific Publishing Tradition

The scientific publishing tradition is a collection of practices and assumptions that have become part of the values and common sense of science. A central tenet of this tradition is publication only after careful and deliberative review by experts. Not only is it considered wasteful to publish a paper that contains errors or repeats earlier work, it is an affront to the tradition of science to publish statements easily refuted by experts. Another tenet is that every published paper is a permanent member of the library of all scientific literature. Many of the scientific societies established their own publishing houses and established review processes; through their membership, they have access to the expert reviewers, and they have a ready-made audience of readers. The societies ensure that repositories exist containing back issues of their publications.

In this tradition, a journal paper passes through four phases, separated by three key moments of public declaration:

- **Preparation**: author drafts preliminary version with early results and obtains informal review by close colleagues. This phase ends with the submission of manuscript to an editor with a request to review and publish it.
- **Review and revision**: editor commissions reviews from several experts, called "referees", and, based on their advice, either rejects or requests revisions from the author. This phase ends with the editor accepting the paper.
- **Publication processing**: editor sends manuscript to publication office for copyediting, layout, queueing, and printing. This phase ends with the actual publication of the paper in a journal and its dissemination to subscribers.

- **Archiving and indexing**: societies and libraries preserve back issues; libraries catalog papers; abstracting services summarize recent papers; citation services accumulate citation indices. Students and other readers use these services to locate works long after they were published.

The second and third phases typically take 6-18 months each, or a total time from submission to publication of 12-36 months. The fourth phase is ongoing. The phases are separated by three key public declarations:

- **Submission**: author declares the paper submitted to an editor; this is documented by a letter to the editor.
- **Acceptance**: editor declares the paper accepted; this is documented by a letter to the author.
- **Publication**: publishing house prints and distributes the copies of the journal issue in which the paper appears.

A copyright transfer usually takes place as part of acceptance. The author grants the publisher the right to use the work in any form for any educational or scientific purpose of the publisher's parent scientific society and retains rights for patents and reuse of the work.

The system relies heavily on the will of the society to continue the journals by marketing and managing subscriptions, setting standards, and appointing new editors. This system also relies heavily on the volunteer efforts of experts and editors. Most of the editorships are volunteer positions; most societies form search committees to locate new editors-in-chief and delegate to the editor-in-chief the authority to appoint associate editors. The reviewers are almost always volunteers; it is the common sense of the field that an author who submits a paper "owes the field" three reviews. In practice, many reviewers report that they receive an average of one manuscript a month for review and that it takes them 2 to 6 months to complete a given review.

Most publishers follow three additional policies. One is a *novel submission* policy, under which an author is expected to submit substantially new material that does not overlap significantly with previous submissions by that or any other author. Second is a *no scooping* policy, under which an author has no authority to distribute copies publicly until the paper has actually been printed. Third is a *proper citations* policy, under which an author is expected to give proper credit to all other persons who contributed to the work in some way, either through previous publication or through private communications. Authors who violate these policies typically receive reprimands from editors and may jeopardize their future right to publish with those journals.

These policies and practices collectively serve to provide an *imprint* or imprimatur to the novelty and soundness of published scientific works. The society gains prestige in the science community by seeking to publish only the most novel, significant, readable, and well-grounded works. The authors gain prestige in the science community by having their works published in prestigious journals. The imprints of a society can be of significant professional value to an author – for example, academic authors consider them essential to promotion and tenure. The harder it is to achieve the imprint and the higher the quality it signifies, the greater its value to an author.

Although less visible, the policies and practices of archiving and indexing are as critical as publishing. A society's imprint would be worthless without reasonable assurances that the published work will be preserved for posterity and that readers can locate the work without having to locate the author. Authors who argue that the publishing process ends with publication are forgetting the importance of archiving to the preservation of their work.

3. Breakdowns in the Traditional System

The traditional scientific publishing system is now facing a variety of breakdowns that must be overcome if the system is to survive. We assume that resolving these breakdowns is preferable to abandoning scientific publishing. From ACM's perspective, the breakdowns are:

1. Most of our journals are written by experts for other experts, but these experts constitute less than 20% of the readership. The other 80%, who are typically experts from other subdisciplines or are practitioners, may be interested in the results but do not have the time or background to understand the specialized language of the journal's domain experts. These 80% are showing their growing dissatisfaction with the enterprise by complaining about too many esoteric papers, dropping their subscriptions and sometimes their memberships, and demanding new kinds of publications they find more approachable. With the increasing penetration of computers into everyday practices of society, this group is growing. In ACM we refer to the traditional line of publications, which are the majority of our journals, as "Track 1". We are gradually introducing a new line of publications aimed at the other readers; we call these the "Track 2" publications. Among other things, the Track 2 publications can bridge between practitioners and research scholars.

2. Authors are increasingly dissatisfied with delays in the process. It often takes 6-18 months to complete the review-revise phase, and another 12-18 months after that until actual publication. Even if we could magically remove the publication delay by clever use of advanced technology, authors would still be dissatisfied with the long review time.

Moreover, readers are dissatisfied if they believe that a result known 1-3 years ago has taken this long to be published.

3. It is an increasingly popular practice among authors to post their manuscripts on publicly accessible file transfer protocol (FTP) servers at or before the moment of submission, thus making the moment of publication precede the moment of acceptance. This practice, sometimes called "circulating preprints", not only accelerates the dissemination of new results, it is seen by many as improving the quality of works by subjecting them to wider scrutiny than that of a few referees. This obviously poses a challenge to the policy of not considering previously published works.

4. New questions are arising about who owns (or should own) the copyrights. Since the FTP server is becoming the author's means of dissemination (at least to a core group of interested persons), some authors now wonder whether there is any value in signing over the right to disseminate to a publisher – and some openly wonder if there is any need for the publisher at all. Other authors are looking to publishers to be their agents in bringing their work to the widest audience, and protecting and preserving their work. Artists, following their standard practice, often retain copyright to their art images, and only give permission to include those images in specific papers; this challenges the policy that the publisher may freely distribute copies of the entire paper and complicates electronic redistribution.

5. Libraries are suffering under reductions of their budgets at a time when subscription prices have been rising faster than inflation and the number of scientific journals has been growing rapidly. They are dropping journal subscriptions and joining together in consortia that share one subscription among several institutions. They do not save all published journals; they look instead to the professional societies to do that. This threatens the archiving function by removing the commitment to retain all works indefinitely. It is highly likely that many scientific works exist as citations only (the original documents have been lost), and that many others have been lost completely.

6. The relentless rise of the number of printed journal papers and their prices, and in the number of manuscripts distributed by electronic means, is causing *information overload*. Individuals and institutions alike are shifting from a mode of acquiring publications for just-in-case use to a mode of acquiring them just-in-time. The latter mode is increasingly facilitated by on-line reference databases and document delivery services. This trend, which appears irreversible, will eventually lead to the disintegration of print journals as preselected collections of

worthy papers.

7. Although publishers say it is not in their mission to cater to academic concerns for recognition, tenure, and promotion, these concerns nonetheless have been a powerful engine in the scientific publishing industry. Authors tend to submit to journals with the highest perceived prestige. Tenure committees are beginning to assess the value of the imprint rather than the print journal itself. Submissions to traditional journals continue to *increase* even as readership decreases, leading to what some are starting to call "write-only journals".

8. Authors are increasingly viewing their works as "living on the web", an allusion to the rapidly growing World-Wide Web (WWW) of interconnected documents. They see networks as new opportunities for collaborative authoring and for *dynamic documents* that incorporate other documents by link rather than by direct copying. Over time, authors want to introduce either new versions or changes into their own works. This is raising new problems of version control, copy-on-demand when exercising a link, reference, citation, and copyright of a nonfixed work.

9. Authors of works stored in the Web increasingly use active hypertext links to other works rather than the traditional citation. Clicking on the link invokes a process that copies the referenced work from a remote site. Such a link, when used, becomes a way of incorporating another work on demand into a document. Link-use is not contemplated in existing copyright policy.

10. Some authors are posting complete collections of their personal works on servers where others can locate them easily simply by knowing the author's name.

In effect, the three key moments of the traditional process – submission, acceptance, and publication – are no longer distinct or in traditional order. The moment of publication is, with the help of public servers, increasingly likely to precede the moment of submission. The moment of acceptance is becoming the moment of imprimatur. Printed publication is becoming less important to authors. The responsibility for archiving and indexing is gradually being abandoned by librarians, who cannot afford comprehensive collections or the software tools for electronic archives.

4. ACM's Response as a Society Publisher

These breakdowns, and the other changes in means of publication and distribution, show the scientific publishing enterprise is being transformed. The broad outlines of what will emerge are already discernible in the practices of some publishers and in the visions many are expressing of the future. These outlines are centered around a structured database containing the society's published works.

- **Journals** will become *streams* flowing into the society's database and will retain their identities as *database categories;* at the moment of acceptance, a paper will be placed in the database rather than into a print queue at the publication house. Separate issues and page limits will disappear.
- **Societies** will offer facilities and mechanisms whereby authors can post collections of their works and obtain public comment on early versions of them.
- **Individuals** will cease to purchase journal subscriptions and will instead purchase a right of access to the entire database. They will post interest-profiles and will be automatically notified when new items matching their profiles are posted. They will read from the database and will be responsible for their own printing. The publisher may provide print copies on demand or by fax for a fee.
- **Publishers** will distribute *notices of availability* rather than journals or documents; readers will locate and obtain copies on demand using new software tools. Local agents specializing in print-on-demand will be established in print shops, copy shops, and libraries, especially at universities.
- **New kinds of services** such as search, extract, and repackaging will be made available.
- **New kinds of works** including hypertext, picture, graphics, sound, and other multimedia effects will be sought and accommodated. New paradigms of works such as training packages will also be accommodated.
- **All interactions** between author and editor, and between editor and reviewer, will be conducted by networked services. This will include all coordination concerning reviews and revisions.
- **The publishers will cooperate** in *virtual libraries,* offering combined access to library patrons. Thus a member of ACM may also have access to works stored by IEEE.
- **Site** licenses granting access to the database will become common because they will be cheaper and more convenient than permissions for individual documents.
- **Copyright release fees** will become nominal and will be collected by automatic meters when documents are extracted from the database.
- **Advertising services** will become more attuned to individual interests and concerns. Links will be established between literature and

related products.
- **The society's database** will also contain non-archival items such as calendars of events, conference schedules, employement opportunities, and industry news.
- **Access** to the society's database and its basic services will be the core of the membership package.

These transformations have already begun. The clock cannot be turned back. ACM authors are already placing documents in databases in the "web" of information servers. ACM has developed enough of a conceptual framework to position itself boldly in the new world whose general outlines are described above. ACM is undertaking to reinvent itself as a publisher.

In response to the shifting digital media and networks, and to the breakdowns enumerated above, the ACM has embarked on a four-part strategy:

1. Move aggressively toward having the entire ACM literature in an on-line digital library. The service capable of supporting capture and production of works should be available by second quarter 1995 and general dissemination within a year thereafter.
2. Ameliorate the problems of Track 1 Transactions by various delay-reducing improvements. Eliminate processing delay by publishing in the digital library. Be prepared to phase out print versions and phase in electronic distributions.
3. Establish a new line of publication offerings (Track 2) for those devoted to using and applying technology, those who seek more general information about current technical developments, and those who seek to understand current research.
4. Engage in many experiments with new forms of publishing and publication services. Add the best of these to the repertoire of digital library services. Revise copyright policies to encourage and accommodate the changes fostered by these experiments. (These strategies are briefly described later.)

Although it is not part of the Publications Plan, the ACM has also made it a top priority to encourage and develop an ACM electronic community, supported by a variety of networked services administered through the Internet host acm.org. All the individuals who volunteer services to the profession through ACM are now using the email and bulletin board facilities of Internet to coordinate their actions. Members who do not have Internet accounts can purchase one from ACM at a nominal monthly fee ($12). The new ACM publication structures will exploit the ACM networked services heavily.

5. ACM Digital Library

The core technology of the ACM approach is a database that serves as an on-line library of ACM's entire literature and offers a range of useful services for electronic publication. It is being developed in two phases. Phase 1 is an initial database and tools to use it for production of publications; this database will come online and begin accumulating content in spring 1995. At that time, all new submissions will be in digital form, and the system will support capturing, storing, and linking certain nontextual objects such as pictures, graphs, equations, sound, or movies. Documents will be stored in several formats, including SGML, that will permit all their component objects to be recognized during searches.

Phase 2 is the deployment of distribution and access services; it includes establishing a network of servers of ACM materials, installing authentication and payment services, developing search and retrieval services, and interfacing with intelligent agent services. In populating the database of Phase 2, ACM plans that works published after 1994 will be stored in full digital format (original and SGML files, and possibly PostScript, PDF, and Lectern format files). Works published before that will be captured whenever possible in digital format, and most works before 1990 will only be available as text images. The database and distribution services are being designed around these assumptions:

- **Electronic documents** whose contents are logically structured for search and retrieval will be preferred to electronic analogs of the printed page.
- **Visualization** of scientific data through multimedia presentations will supplement and enhance text-only documents.
- **Documents** will be object-oriented, with some components being other objects already published on the Web.
- **Interactive Documents** will supplement read-only documents.
- **Document Delivery Just-in-Time** will require easy access, high availability, and good performance.
- **Access** from home, work, or school desktops from around the world will become a primary mode of acquiring knowledge. Good network access is essential.
- **Tools** to help avoid information overload will acquire central prominence. These will include personal information agents to assist users in selection, filtering, and interpretation. They will include standard interfaces to distributed collections of scientific information from many societies, and will replace the current eclectic set of Internet tools and protocols. Societies will continue to serve domain and discipline needs.

We have made it a high priority to develop authentication services which will be needed to control access to the database and its functions. We will implement new functions, notably access licenses for institutions, short-term licenses for nonmembers, promotional licenses, and triggered functions such as intelligent agents that collect copyright release fees from nonmembers accessing ACM copyright works.

6. Track 1 Publications

Our short-term objectives are that all print journals and Transactions be published on their schedules; that some be expanded from quarterly to bi-monthly publication when the backlog and subscribership would support the increased capacity; that joint journals in overlapping interest areas be established with other societies, for example, the *Transactions on Networking* with IEEE; that specialized journals of other publishers be offered at good prices to our members, for example, the *Multimedia Systems* journal of Springer-Verlag.

Our long-term objective is to transition all our journals to on-line distribution. A number of benefits arise from this: articles are available sooner; costs of printing and binding can be shifted to local sites where they become optional; postage and warehousing costs can be eliminated; individuals can gain access to articles without subscribing to a whole journal; and preliminary versions of papers can be posted for public comment. Print versions will be phased out as the demand for them becomes too small, an outcome that may happen for some journals as early as 1998. We do not expect that print versions of Track 1 publications will be a major source of revenue for ACM in the long term.

We are also undertaking experiments in electronic distribution. In 1995, a new on-line electronic journal of combinatorial and numerical algorithms will begin operation; it will include the current *Collected Algorithms from ACM* (CALGO). Subscribers to *ACM Transactions on Database Systems* (TODS) will be offered on-line access to the queue of TODS papers that have not yet been published. Subscribers to *ACM Transactions on Programming Languages and Systems* (TOPLAS) will be offered on-line access to appendices of published papers, which can then be printed in a shorter form. Conference proceedings will be distributed on CD-ROM—Multimedia '93 and '94 and Supercomputing '94 are early examples. Back volumes of journals will be offered on CD-ROM.

7. Track 2 Publications

Many ACM members have expressed concerns for learning and effectively using the best new results of technology. ACM has responded to these concerns by repositioning the *Communications* , by undertaking a new line of Track 2 publications, and by cooperating with some commercial publishers on offerings for our members' Track 2 interests. ACM started two new Track 2 publications in 1994: *StandardView* , a magazine devoted to the debates and controversies in the field of standards, and *interactions* , a magazine devoted to exploring and elucidating new and varied ways computing can reach the world.

We expect print versions of Track 2 publications, including those on CD-ROM, to be a viable business: not only is the market for them wider, but their preparation tends to be sufficiently expensive and time-consuming that most authors will seek professional help for their production and will expect income from their use.

Track 2 is not just a type of publication, it is a way of thinking about engaging researchers, developers. and practitioners together in the ongoing professional learning process. It is a new way of generating offers for members. Other parts of ACM. such as SIGs and education, are also considering new, Track 2 programs.

8. Experiments

Experimentation with new practices is the only way to find out which ones will be effective. Accordingly. we encourage experiments in electronic publication and seek to facilitate them with new copyright policies. Here is a partial list of the experiments that are underway or will be undertaken soon:

- **Online** journal of algorithms (including combinatorics and CALGO)
- **Subscriber** access to backlog queue of TODS
- **Publishing** TOPLAS appendices online, thereby shortening print papers
- **Online** payment and authentication systems
- **Local** (e.g. campus) distribution in return for digitizing past literature
- **SIG** conference proceedings on CD-ROM or server
- **Back issues** of journals on CD-ROM
- **Participation** in Stanford NSF/ARPA/NASA-sponsored digital libraries project
- **Participation** in *Journal of Universal Computer Science* (JUCS), a multinational publication venture in the WWW

- **Cooperation** with MIT Press in distributing *Chicago Journal of Theoretical Computer Science*
- **Metering** use to charge for copyright release
- **Local agents** (e.g. libraries) for search and print-on-demand
- **CD-ROM** distribution of images of printed materials in highly readable formats

9. New Services

The structured database described here positions ACM to offer new services that will make ACM members differentially more competitive than nonmembers. Over time, ACM expects to realize less revenue from print media and Track 1 publications and more from three new principal businesses:

- *Guided access to literature:* Members will be given access to the ACM digital library (and to similar services of cooperating societies) from which they can search and extract documents or summaries. They will be notified of new items that match their interest profiles. Nonmembers can purchase short-term licenses.
- *Conferences:* Conferences will continue to expand. Some of them will be conducted in the Internet. Proceedings will be rapidly available either by network or by CD-ROM.
- *Continuing education:* ACM will offer reading and discussion programs based on collections from the database. Those who pass the quizzes designed with these programs will receive certificates of knowledgeability issued by ACM.

10. Servers and Links

It is becoming a standard practice among engineers and scientists to post copies of their papers on servers attached to the Internet and maintained by their employers. These papers can then be accessed from other servers in the Internet and copied by some form of FTP. Readers can attach comments to the posted versions, and authors can post revised copies. Some editors have established moderated preprint comment services to assist authors and to guarantee that no papers or comments can be modified once posted. These practices are widely seen among authors as means to speed the distribution of findings and to improve the quality of papers and algorithms.

A growing number of professional authors and researchers are posting complete libraries of their personal works on servers; they seek protocols whereby the server holding the complete works of any author can easily be located. This is seen by many authors as a way of establishing a network identity and making their works more readily available to anyone who wants them.

It is also becoming a strandard practice to think of papers as collections of objects (sections, paragraphs, figures, tables, pictures, and the like) rather than simply as texts. The WWW offers the technology of links, allowing authors to embed pointers to, rather than copies of, objects in their works; the reader can click on a link and thereby invoke a process that calls a copy of the object to the local computer.

The new practice of links-use is widely seen by authors as a means to constructing multimedia, nonlinear documents that incorporate by reference relevant works from anywhere in the world. It is also seen as a way to simplify construction of new works that rely on other works: the author of a work does not have to obtain prior permission to include another work since the other work is not actually incorporated at the time of writing. In other words, the link is seen as a citation, and a copy of the work is obtained upon an explicit request by the reader.

These new practices are bringing authors and readers into conflict over copyright laws. Authors maintain that links are citations and it is the responsibility of the copyright owner to demand permission when a reader uses the link. Copyright holders maintain that the author is, in reality, intending to make a copy available to the reader and must obtain prior permission. Copyright holders are beginning to design authentication servers so that certain people (such as members of a professional society) can get free access as part of their dues while others must pay to gain access; the holder may offer the prospective reader a free preview to help that reader decide whether a full copy is worth paying for.

ACM has decided to treat links as citations. ACM encourages wide use of links as citations. Authors will not have to seek prior permission to place links to ACM copyright works in their new documents. A reader who decides to use a link will negotiate access with ACM at the time of link-use, and ACM will provide mechanisms to make this simple. ACM members and authors will not be subject to copyright release fees when fetching from the ACM databases.

The scientific publishers, such as ACM, are examining ways to structure their copyright policies so they can preserve the spirit of the existing copyright laws within the context of new practices for using servers and links. Until people have settled into standard routines with the new practices, authors and readers will have to think carefully about the copyright implications of their actions.

11. Policy Questions

The experiments and new media are shifting traditional practices, demanding new policies to cover all aspects of the transformed publications processes. The foregoing discussion reveals a number of policy questions that were not contemplated when the existing policies were formulated.

1. Who holds what rights? Do traditional copyright principles apply to digital versions and transmissions? What rights do authors retain? Their employers?
2. What rights do authors and ACM obtain for object-oriented documents, some of whose components are already-existing, published objects referenced by active hypertext links in the Web? What happens when an author obtains permission, but not copyright, for an object belonging to another author? What are the rules for fetching a copy of an object by exercising a link?
3. Does the new, emerging practice of posting submitted manuscripts on public servers constitute publication? Under what circumstances should ACM retain its no scooping policy? What about its novel submission policy?
4. What notices should authors of submitted papers be required to include with their public-server postings? How should an author's personal copies be treated after copyright is transferred to ACM?
5. Should changes and corrections create new versions of a work rather than replacing old versions? Will articles become more like software, requiring management by version control systems?
6. Should high-quality conferences with outstanding reputations be considered of equivalent quality to Transactions and journals? (Conferences review for accept or reject under strict deadlines while journals review for revisions that will improve a manuscript; do these differences matter?)
7. Should there be an archiving fee, replacing the current practice of page charges? Should uncited items be deleted from the archive after a minimum holding time? Should highly cited items be guaranteed a permanent place in the archive?

Answers to these questions are evolving as the field changes and we learn more.

Publishers that learn to provide well-structured knowledge through digital libraries and easy-to-use tools will be the main survivors and successful entrepreneurs in the new medium. They will need to develop new policies consistent with their evolving practices and their long-term vision.

THE ACM INTERIM COPYRIGHT POLICY

Version 2 – Issued by the ACM Publications Board 11/15/95

P.J. DENNING
George Mason University,
Fairfax VA 22030, USA
pjd@cne.gmu.edu
http://www.cs.gmu.edu/faculty/denning.html

AND

B.A. ROUS
Association for Computing Machinery, Inc.
1515 Broadway
New York NY 10036-9998, USA
rous@acm.org[†]

1. Background

Since 1994, ACM has been preparing to shift all its publication operations from paper-only to electronic distribution from a structured database. The first-phase computer system to support this will begin receiving author manuscripts in early 1996 and will store them in SGML format from which printed versions and electronic versions can be generated. ACM is doing this to accommodate shifts in author and reader practices that are emerging in the world-wide Internet. With this copyright policy, ACM aligns itself with those practices and becomes the first scientific publisher to adopt copyright policies for cyberspace.

By the end of the decade, we envisage a world of scientific and technical publishing with three main characteristics. First, the entire technical literature of a field will be stored in a digital library, a network of databases offering new kinds of services such as browsing, searching, extracting, and repackaging; simple pricing schemes will be used to collect nominal fees from those who have not subscribed to the database services; the library will inform subscribers when new material of interest to them has been

[†] © Copyright 1995 by ACM, Inc. Permission to copy and distribute this document is hereby granted provided that this notice is retained on all copies and that copies are not altered.

Astrophysics and Space Science **247**: 83–93, 1997.
© 1997 *Kluwer Academic Publishers.*

posted. Second, works will be stored definitively on servers warranted and maintained by copyright holders as a service to authors and readers. Works may evolve as authors and editors agree on changes. Third, copyright permissions policies will be adapted to the realities of electronic dissemination by encouraging unlimited use of hypertext links, leaving the payment of any fees to on-the-fly negotiation between reader and copyright holder, and permitting virtual publications in which individuals assemble readable views of documents from protoype documents containing links.

This policy is based on five principles.

- **Publishing works of quality.** ACM intends to retain its reputation as a publisher of materials worth reading. ACM will review or edit all submitted or solicited items to certify that they meet ACM's high standards for quality and reliability.
- **Maintaining integrity of works by ACM authors.** ACM will warrant that the copies of works in its digital library are definitive: that they have not been modified or altered without author and editor permission. ACM will be able to provide proof of these warrants. ACM will store the definitive versions and give unlimited permission to copy links to those versions. (The ACM web page contains a figure illustrating the relations among these versions.)
- **Transcopyright permission for electronic dissemination.** ACM incorporates a principle similar to one named "transcopyright" by Ted Nelson. ACM will hold its copyrighted works on its servers and will give free and unlimited permission to create and copy links to those works or their components. So that readers can locate the context from which an excerpt was drawn, ACM will provide a way of linking a component to its parent work. Readers following links will gain access upon payment of a fee or presentation of a valid authorization certificate to ACM or ACM's agent; ACM or its agent will issue a personalized certificate of ownership to that reader. A person owning a copy may not replicate that copy and give it to others unless the copy carries explicit permission for further replication.
- **Author-friendliness.** ACM intends to be the author's agent in reaching the widest possible readerhip and protecting the author's interests against plagiarism and unauthorized copying or attribution of an author's work. The ACM grants authors liberal retained rights including unlimited reuse of the work with citation of the ACM publication and the right to post preprints and revisions on a personal server. ACM will take legal action against those who infringe its copyrights.
- **Emphasis on value-added services.** ACM is a member organization chartered to disseminate information about computing broadly to its

members and to the public. ACM will assist readers to locate materials of value to them. ACM treats copyright ownership as a means to allow it to provide a digital library to its members and the public and to act against anyone attempting to duplicate ACM's library; ACM does not treat copyright permissions as a significant source of revenue.

These new policies clarify the liberal conditions under which ACM grants permission for copying or distribution, and the conditions under which ACM requires prior permission and/or a fee. A glossary of the principal terms is included at the end.

This statement of policies is marked as "interim" because the Publications Board expects to learn from experience how effective the various provisions are. The Board will conduct a review of these policies annually, revising them as needed to deal with new circumstances and to accommodate innovations. This document supersedes all previous statements of ACM copyright policies.

2. Copyrighted Works

2.1. REQUIREMENT FOR COPYRIGHT

ACM asks authors to assign copyright to ACM as a condition of publishing the work with ACM. This requirement may be waived for materials that have not been reviewed or refereed. Immediately after the copyright transfer, authors should incorporate the ACM copyright notice into their personal copies.

Authors of new works who wish to embed a copy of (not just a link to) a component of an ACM copyrighted work, e.g. a table or a figure, must obtain explicit permission of ACM. (See also "Permissions" Section 2.3.

An author who embeds an object, such as an art image, copyrighted by a third party is expected to obtain that party's permission to include the object with the understanding that the entire work may be distributed as a unit to ACM members and to others in any medium. The copyright transfer applies only to the author's work as a whole, and not to the third party's embedded object. The requirement to obtain third-party permission does not apply if the author places only a link to the copyright holder's definitive version of the object.

2.2. COPYRIGHT NOTICE

The ACM copyright notice must be displayed on the first page or initial screen of a display of all works copyrighted by ACM, whether those works are published in print or in a digital medium. It is acceptable to place the string "(c) Copyright 199x by ACM, Inc." as a hypertext link to the full copyright notice.

2.3. PERMISSIONS

A person granted permission to copy an ACM work should display with the copy (a) the notice "Included here by permission, (c) ACM, Inc." and (b) a link or citation to ACM's definitive version. The link or citation will enable a reader to access the context in which the copied material originally appeared. Full copies of the work should also include the full copyright notice, which will normally be a part of the work anyway.

ACM publications staff will monitor permissions@acm.org for requests for permissions and releases under this policy.

2.4. INTERIM PERMISSION TO MAINTAIN DEFINITIVE WORKS

Until the ACM database is operational, authors are granted permission to maintain one copy of the ACM definitive version, for which they have transferred copyright to ACM, on a non-ACM server.

2.5. DEFINITIVE VERSIONS AND REVISIONS

ACM will create and maintain a definitive version of every copyrighted work. Definitive means that ACM certifies that this is the work approved by the author and editor and has not been changed since the last version. The definitive version may differ from an accepted version because it has been edited by ACM. A given work, such as a dynamic book, may evolve through several definitive versions as authors and editors approve and incorporate changes. ACM will create conventions for citations to specific versions.

As part of their retained rights, authors may make changes to their ACM copyrighted work and post the changed version on a non-ACM server. If the changed work differs by 25% or more from the copyrighted version, it is treated as a new work not copyrighted by ACM; otherwise it is treated as a revision and is still copyrighted by ACM. A revision should be marked as such and should include a citation and link to the definitive version. (See also Section 3.2.)

ACM asks that authors exclude from their personal collections copies of the definitive versions maintained by ACM, using instead links to the definitive versions. (See also Section 3.2.)

2.6. FIXITY OF WORKS

The electronic media provide means whereby readers can attach comments to an author's work and the author can respond. The ACM wishes to encourage this and intends eventually to support this as a service in the ACM digital library.

ACM subscribes to the general scientific convention that published works not be altered without review and approval by an editor. ACM also considers that all reader and author comments formally attached to a work are part of the public discussion and should not be altered by their authors without approval by an editor. If the author or a reader wishes to withdraw a comment after posting, the withdrawn item will be replaced in the public record by a withdrawal notice; ACM will retain a private copy of the withdrawn item.

2.7. INTERPRETATION OF COVERAGE

ACM has a long-standing policy that the copyright transfer statement grants ACM the right "to publish the work in whole or in part in any and all media". ACM has always interpreted this policy to include digital media, digitized copies of previous print versions, performance and display by reading, and digital transmission of files containing the copyright works. ACM hereby reaffirms this interpretation.

3. Rights Retained by Authors

3.1. AUTHOR RETAINED RIGHTS

As part of a copyright transfer to ACM, the original copyright holder (author or author's employer) retains:

- All other proprietary rights to the work such as patent,
- The right to reuse any portion of the work, without fee, in future works of the author's own, provided that the ACM citation and notice of the ACM copyright are included, and

– The right to post personal copies of preprints or revisions in a personal collection on a non-ACM server; these copies must be limited to noncommercial distributions and personal use by others, the ACM copyright notice must be attached, and the server must display a general policy notice about the presence of copyright works it contains.

3.2. AUTHOR PERSONAL VERSIONS

A link to the author's "personal copy of paper" should take a reader to a page in the author's personal collection containing links to the ACM definitive version and to any other versions, such as preprints and revisions, maintained by the author. All the versions beginning with the one accepted for publication by ACM should bear the ACM copyright notice. Until the ACM offers its digital library service, authors may also store the definitive version on the same server.

4. Processes

4.1. OPEN NOTICE OF SUBMISSION FOR PUBLICATION

An author who submits a work for consideration by an ACM editor should include this notice on any personal copies posted on servers:

> This work has been submitted for publication. Copyright may be transferred without further notice and the accepted version may then be posted by the publisher.

ACM and other publishers have a policy that authors submit a work for consideration for publication by only one editor at a time. Authors must notify editors if a work is identical or substantially the same as another work submitted or accepted for publication.

4.2. REPUBLICATION

ACM maintains its policy of not republishing works, whether copyrighted by ACM or by others, except under limited conditions where an editor determines there is significant benefit in republication.

4.3. EDITED COLLECTIONS

In most cases of conference proceedings, newsletters, and other edited collections, the collection as a whole and all its components will be copyrighted by ACM solely or jointly with other organizations. In some cases, notably newsletters, the collection will be copyrighted but copyrights of some components will be retained by authors.

No collection in which ACM is the sole or joint copyright holder may be posted for open distribution without prior permission from ACM. Notice of permission must accompany the ACM copyright notice. Free access may be granted to conference attendees and appropriate groups of ACM members provided an authentication mechanism is in place.

4.4. SOLICITED WORKS

From time to time, ACM solicits works for publication. Examples are columns, invited articles, award lectures, and keynote speeches. ACM asks authors of such works not to distribute copies or post these works until ACM has published them. Authors who wish to circulate before publication should get permission from ACM. ACM considers lectures and speeches to be published at the time they are given.

4.5. ELECTRONIC PUBLICATION EXPERIMENTS

SIGs and other units of ACM are encouraged to conduct experiments in electronic publication and distribution provided that the experiments conform with all the policies stated here and prior notice and description are given to the ACM Director of Publications. Quarterly progress reports should be sent to the ACM Director of Publications for the duration of the experiment.

5. Access to Copyrighted Works

5.1. ACCESS LICENSES

ACM will provide all ACM members in good standing with a license to access the ACM database and its basic services as part of the regular membership package.

ACM will offer licenses to others for access to ACM publications databases for purposes such as access, searching, extracting, or downloading. Licenses that allow print-on-demand may include a per-copy release fee.

Institutional members of ACM may obtain licenses to download items from ACM databases for internal redistribution upon certifying they have authentication services capable of limiting redistribution to their members.

ACM will also offer limited-time access licenses to nonmembers. Such licenses can be used as promotions for ACM membership as well as allowing someone an opportunity to use ACM published works for a limited time.

5.2. LINKS

A link is a string that, when interpreted by an appropriate program, will reference an object and make a copy accessible locally. Examples are hypertext links, URLs (uniform resource locators on the World-Wide Web), and document handles. ACM treats links as citations (references to objects) rather than as incorporations (embeddings of objects).

Permission to access, and payment of applicable fees, are matters of negotiation between a reader who exercises a link and the rights-holder for the referenced object. ACM encourages the widespread distribution of links to the definitive versions of ACM copyrighted works and does not require that authors obtain prior permission to include such links in their new works.

If an author wishes to embed a copyrighted object rather than a link in a new work, that author needs to obtain the copyright holder's permission. (See also "Permissions", Section 2.3.

Someone who creates a work whose pattern of links substantially duplicates a copyrighted work should get prior permission from the copyright holder. For example, the creator of "A Table of Contents for the Current Issue of TODS" – consisting of citations and active links to authors' personal copies of the articles in the latest issue of TODS – needs ACM permission because that creator is reproducing an ACM copyrighted work. If all the links in the "Table of Contents" pointed to the ACM definitive versions, ACM would normally give permission because then the new work advertises an ACM work. To avoid misunderstandings, consult with ACM before duplicating an ACM work with links.

Service providers do not need to obtain prior permission from ACM to locate and dispense *links* to the ACM definitive versions of works, but they do need permission if they are making, collecting, or distributing *copies* of ACM copyrighted works.

ACM intends to offer receipts of ownership to those who obtain authorized digital copies of ACM works so that they may certify the validity of their copies. ACM intends also to make available links to components of its published works (e.g. tables and figures). These links will allow the components to be accessed in their original contexts.

5.3. DISTRIBUTIONS FROM NON-ACM SERVERS

Individuals often distribute copies of works authored by themselves or by others. Distribution may consist of sending copies to a mailing list or of posting a copy on a server where it is accessible to others who might copy it. Electronic distributions and postings of ACM copyrighted works are acts of copying and may require ACM permission.

Authors who have transferred copyright to ACM may post copies of preprints and revisions in their personal collections as specified in Section 3.2. With ACM permission, their employers may offer a copy of the definitive version for use within the organization.

Anyone who legitimately obtains a copy of an ACM copyrighted work may use the copy only for non-commercial classroom or personal use, as specified in the ACM copyright notice.

If ACM copyrighted works are maintained and distributed from non-ACM servers, ACM requires that the server prominently display a general notice alerting browsers to the presence of copyright materials. A sample of an acceptable notice is shown below.

Sample of Server Notice:

> The documents distributed by this server have been provided by the contributing authors as a means to ensure timely dissemination of scholarly and technical work on a noncommercial basis. Copyright and all rights therein are maintained by the authors or by other copyright holders, notwithstanding that they have offered their works here electronically. It is understood that all persons copying this information will adhere to the terms and constraints invoked by each author's copyright. These works may not be reposted without the explicit permission of the copyright holder.

5.4. PRODUCTION OF DIGITIZED COPIES

Persons who have permission under these policies to make copies may elect to digitize a print copy and distribute the digitized copy. Because digitizing processes such as OCR (optical character recognition) are error-prone, this disclaimer must be included with the ACM copyright notice on each digitized copy:

> This is a digitized copy derived from an ACM copyrighted work. ACM did not prepare this copy and does not guarantee that is it an accurate copy of the author's original work.

5.5. ELECTRONIC RESERVES AND COURSEPACKS

Schools, colleges, and universities and other nonprofit educational organizations may place digital copies of ACM works in their library's electronic reserves for the duration of a course that uses those works.

Anyone manufacturing coursepacks commercially must pay a fee to include ACM copyrighted materials. This can be covered through blanket licenses with the Copyright Clearance Center (CCC) or directly from ACM.

6. Definitions

Some of the words in this policy have specific meanings in ACM's domain. The meanings intended herein are recorded follows:

- **Work**: A document, file, manuscript, or other information object, in any form, that is an expression by an author protected under copyright law. Works will be stored in the ACM database (normally in the SGML format). Browsers and viewers make local copies that are rendered for display; these copies are considered personal copies of the person using the browser.
- **Definitive version of work**: A version that has been accepted by an editor after review, which may have been professionally edited, and which contains the full citation and ACM copyright notice. The definitive version is generally different from the version accepted by the editor. A work may evolve through a sequence of definitive version as authors and editors agree on further changes. Definitive versions will be protected from unauthorized alteration.
- **Personal Copy, Author**: A work maintained by the author. There may be several versions including the preprint version, accepted version, and revised version(s). The definitive version is not considered part of the author's personal version-set and is maintained by ACM.
- **Personal Copy, Reader**: A version constructed by a reader that has replaced links with their referenced objects after negotiation with the copyright holder; or a printed copy of this version. It is not is not intended for further replication unless the copyright holder gives explicit permission.
- **Edited**: a collection of works have been selected by an editor and possibly edited for style and length.
- **Reviewed**: one or more experts have examined the work and have given assessments to an editor about clarity, soundness, novelty, prior publication, proper citations, and other criteria.
- **Formally reviewed**: A thorough review with emphasis on clarity, accessibility to the general reader, and timeliness. Persons serving as formal reviewers are independent of the editors who request their advice.
- **Refereed**: A thorough review with emphasis on novelty and soundness. A journal refereeing process seeks to advise the editor whether to reject or provide specific guidance for revisions. A conference refereeing process seeks to advise the editor whether to accept or reject; a strict deadline is enforced. Persons serving as referees are independent of the editors who request their advice.

- **Journal, Transactions**: generic names given to ACM refereed periodical publications.
- **Communications, Surveys, Interactions, StandardView**: four ACM formally reviewed periodical publications.
- **Proceedings**: the reviewed or refereed record of a conference.
- **Newsletters, Bulletins**: edited and/or reviewed periodic publications that inform members of groups about relevant news.
- **Link**: A character string that denotes a work stored at a remote location in a network; the link is associated with a protocol for retrieving a copy of the item denoted by the character string. Invoking or exercising a link means to call a function in the protocol that fetches a copy of the work into the local computer.
- **Server**: a computer in a network that stores files and databases of works and provides means to access and copy those works to other computers.
- **ACM database or digital library**: the entire collection of ACM copyright works and associated services. It may be stored on one or more machines.

THE ICSU PRESS PROGRAMME
ON ELECTRONIC PUBLISHING IN SCIENCE

D.F. SHAW
Keble College
Parks Road
Oxford OX1 3PG, UK
dennis.shaw@keble.ox.ac.uk

1. Introduction

Scientists rely on accurate, rapid and widely disseminated publication for the communication of the results of scientific investigation. ICSU Press was established to ensure that these needs were satisfied for all scientists regardless of discipline or geographical or political factors. With the development of the Internet, transmission of digitally stored texts and graphics added a new dimension to the process of scientific communication. This process is the subject of this chapter on electronic publishing. ICSU Press now provides an advisory service on publishing for the members of the International Council of Scientific Unions (ICSU). In 1993 the author proposed that a major study of electronic publishing should be carried out to identify issues of concern and to advise on future action for the benefit of science. Authors and publishers recognised that electronic publishing would materially affect the important role of editors in managing the publication process; and many of them thought there was a serious threat of a breakdown in the long established system of publication through scientific journals due to loss of revenue required to support the system. A preliminary study identified two issues requiring special attention before the intellectual property rights in the digitally stored texts of research results could be fully exploited. These were: (1) that there was a need for an internationally recognised system for the bibliographical control of digitally stored texts; which was a matter of concern to publishers as well as authors; and (2) that a reliable and economical method was required for the protection of digitally stored texts against non-permitted copying.

Astrophysics and Space Science **247**: 95–116, 1997.

2. A Meeting of Experts

To explore these and related matters, representatives of international organisations, who were interested in bibliographic control and protection of intellectual property rights in digitally stored texts, were invited to a small gathering of experts held in collaboration with UNESCO in Paris in June 1993. The meeting was invited to consider what initiatives should be undertaken by ICSU in seeking to satisfy these needs.

A report of this meeting [1] was presented to the 24th General Assembly of ICSU held in Santiago (Chile) during October 1993. The report was received with enthusiasm and the Assembly resolved that ICSU Press, in close collaboration with the Scientific Unions should maintain a watching brief on the rapid expansion of electronic publishing which was expected to occur during the next few years and which may (1) seriously affect the economic viability of journals publishing, (2) engender a reduction in the use of the peer review system and (3) result in an increase in the problems concerning intellectual property. Recognising the maintenance of the necessary infrastructure to be of the utmost importance to authors, publishers and readers as these developments take place, ICSU Press was requested to report periodically to the ICSU Executive Board to ensure that the best interests of the international scientific community were taken into consideration.

3. An International Conference of Experts

In order to respond to this invitation it was clearly important to ensure that progress was carefully monitored and also that we were fully informed on the latest issues. It was decided to convene a second larger conference of experts to be planned with the assistance of an international advisory committee the membership of which was chosen in order to complement the expertise of members of the Editorial Board (see Section 8).

3.1. CONFERENCE PLANNING

Planning for the conference was the responsibility of ICSU Press which, in collaboration with UNESCO, established a joint Programme Committee to work out the detailed programme. Corresponding members were invited to join this committee to form the Editorial Board that was responsible for the Proceedings. To ensure a common core of knowledge for all participants, speakers were chosen to review the major topics and issues detailed below, after which delegates were to be invited to participate in workshops for the discussion of key questions identified in the main sessions. It was expected that the responses to these questions would enable recommendations to be drafted for discussion and, if approved by the Conference they would be presented to the 25th General Assembly of ICSU and reported to the Director-General of UNESCO. This plan was adopted and came to fruition in February 1996.

3.2. PLENARY SESSIONS

The subjects selected for the plenary sessions were the following:

1. Introduction-where are we now?
2. Electronic data storage, access and archiving;
3. Tools and standards for protection control and presentation of data;
4. Legal and ethical issues in electronic publishing;
5. Scientists' views of electronic publishing and issues raised, & Views of learned and professional scientific society publishers;
6. Economics and organisation of primary electronic publishing; and
7. Options for the future.

Rather than issuing a call for papers the members of the Programme Committee selected experts with established views on each of these topics to prepare and present summaries. These formed the basis for the drafting of questions to guide the Working Groups in their discussion of the important issues. In some sessions there was provision for short contributions by delegates from countries in the process of economic and social development. Comments on the outline plan were sought from members of the International Advisory Committee. The papers presented in the plenary sessions are published in the proceedings [2,8] and summarised by Shaw [3].

3.3. SELECTION OF PARTICIPANTS

The success of the workshop plan was dependent on achieving a satisfactory representation of participants from research scientists, publishers, librarians and information brokers. It was also important to attract scientists from developing countries. Delegates were nominated by ICSU Scientific Unions, Commissions and Committees, National Members and Scientific Associates; and nominations were also sought from UNESCO delegations. Invitations to those selected to participate were issued by ICSU Press and UNESCO in September 1995. Others, particularly publishers and information brokers with an interest in the conference topic, were invited to apply for a restricted number of places which were allocated when the main enrolment had been completed during the autumn 1995. As a result, a conference of 150+ participants was achieved with the composition shown in Table 1

below. The distribution of participants by country of origin shows a strong
contingency from Europe and North America. There was a bias in favour of
physics and related disciplines which was evidence of the progress in electronic publishing achieved by the International Union of Pure and Applied
Physics and its member societies, together with other ICSU Unions such as
the International Astronomical Union and the International Union of Crystallography. However, the fact that forty countries succeeded in nominating
expert representatives, shows that international coverage for this Conference was achieved. It was generally agreed that the method of selection
produced a valid world-wide conference of experts.

TABLE 1. Country of origin for Conference participants

Country	Number	Country	Number	Country	Number
Australia	1	Italy	1	Poland	2
Austria	2	Ivory Coast	1	Russia	3
Belgium	2	Japan	1	Slovakia	1
Brazil	1	Jordan	1	Sweden	5
Canada	3	Lebanon	1	Spain	1
Denmark	4	Lithuania	1	Switzerland	2
Egypt	2	Luxembourg	1	Taiwan	1
Eire	1	Malawi	1	Tunisia	1
Finland	3	Mexico	2	Turkey	1
France	13	Morocco	1	Uganda	1
Germany	6	Nepal	1	UK	43
India	3	Netherland	10	US	23
Iran	1	Nigeria	1	Venezuela	1
Israel	1	Norway	2	TOTAL	153

3.4. PLANNING FOR THE WORKING GROUPS

The *modus operandi* for the working groups, was determined by the Programme Committee two months in advance of the conference. At that time
the composition of the conference body, in terms of discipline, interests, and
experience, was clear. Thus, decisions could be made on such matters as the
probable size of the groups, and how many questions should be addressed
to each of them. Speakers were asked to provide long abstracts (circa 500
words) on their topics, followed by a first draft text of their papers including
an indication of the vital issues they foresaw as arising. It was thus possible
to identify several of the matters to be explored in discussion. Participants

were notified of the broad subject areas selected for the five groups and were invited to express a preference.

4. Working Group Reports

There were five working groups, viz.:

1. Real Costs (Anthony Watkinson, Chairperson),
2. The Electronic Archive (Robert Wedgeworth, Chairperson),
3. Developing Countries (Ana Maria Cetto, Chairperson),
4. Implications for Training and Work-styles (John Rose, Chairperson) and
5. Peer Review (Bernard Donovan, Chairperson).

Several new issues were brought up in the working group discussions and resulted in a wide ranging set of recommendations. These recommendations, were aimed initially at ICSU and UNESCO but must be addressed by a wider community of scientists and publishers. They represent the most important result of the conference and have been widely distributed. For ease of reference they are included in an appendix. To implement these recommendations it was necessary to attract the attention and supporting skills of each participant and to achieve this end required the maximum publicity. Each delegate was asked to publicise the results and recommendations as widely as possible and this has been done by many.

4.1. WORKING GROUP 1: REAL COSTS

Group 1 was mainly composed of people with publishing responsibilities but most of them were scientists and many had affiliations with learned societies. The group concentrated its discussions on the costs of primary publishing in print form in serials, on the costs incurred in making the print journal electronically available and on the savings resulting from the same process. Secondary publishing, publishing in non-serials and e-journals (journals with no print equivalent) with their special costs were only discussed tangentially. The group did not like the term *costs* and preferred *investment* or *economics* as the circumstances of the discussion allowed, but this preference was not consistently adhered to.

Two publishers, Blackwells Science (Campbell [4]) and the American Physical Society (Lustig [5]) shared with the group their rather different analyses of the costs of producing a journal. Many of the publishers present did not accept the validity of either of the ways in which costs were presented and several others did not agree that it was possible to separate the costs in this way. This was because they represented only an arbitrary slice at a particular point in time of the investment needed to transform the whole publishing process in line with the sort of developments that were discussed at the Conference. Some also felt strongly that it was impossible to compare print and electronic publishing in this way when their thinking was in terms of creating a common electronic platform which would enable the product to be run off in a number of different formats. The only area of agreement was the high percentage of fixed costs (in the range of 70 to 80%) which are incurred before any copies of any journal are printed or distributed. Because assumptions about costs are so central to the debate about the impact of electronic publishing particularly on the libraries, Recommendation III.3 was submitted.

It was recognised that the economics of the whole STM publishing process is underpinned by the willingness of the academic community to work for less than consultancy fees and often for nothing. These hidden costs were not explored. There is no reason to suppose that the situation as existing will change with electronic publishing. There was a long discussion about direct costs to authors of the publication process, particularly because of the model, presented by Joshua Lederberg among others, under the terms of which the author pays for submitting the paper while the reader receives the paper free. This model did not find favour with the group. The perception was that the culture of authors paying to publish through page charges or submission charges is neither in good shape in the USA where it has been common nor has it ever been the preferred way of operating outside that country. In addition there was strong resistance to what was perceived as a tendency for publishers to push back to authors the costs of formatting electronic files. Nevertheless it was felt that, as the dissemination of the results of research is an intrinsic part of the research itself, a resolution could appropriately be put forward recommending the extension throughout the world of the practice common among government and non-governmental funding agencies in the USA of taking into account publication costs in research funding [Recommendation III.1].

It was also the view of the group that investment by the publishing community in electronic dissemination was pointless if the advantages of electronic access cannot be realised because the journals cannot be brought to the desk of the individual researcher due to a lack of internal networking and training. It was felt that funding should be earmarked especially for the building up of an appropriate infrastructure of this sort within the context of library provision [Recommendation III.2].

The group was also concerned with the infrastructure in a wider sense with particular concentration on concerns that the scientific community will not have available sufficient band- width at a lower price for communication over the Internet to be made easier and more effective rather than becoming restricted and more expensive. The concerns of this group are partly reflected by a resolution derived from the deliberations of another group [Recommendation IV.2].

4.2. WORKING GROUP 2: THE ELECTRONIC ARCHIVE

New electronic technologies are continuing to alter the way science is done, the way science is taught and the way science is applied to the various aspects of our lives. There is little surprise, therefore, that these same technologies are continuing to alter how we document the record of scientific achievement and how we give access to that record. This group devoted a considerable amount of time to a number of perspectives on the electronic archive. While there is general agreement on the need to ensure that a permanent record of scientific achievement will be maintained, there were varying levels of agreement on the nature of the electronic archive; what its purposes are; how accessible it should be; who is responsible for maintaining it and who will finance it.

There is some difference of opinion as to whether an electronic archive should be a static collection (or network of collections) as distinct from an active record with assurances of the preservation of the file. There was even less agreement as to who should be responsible for ensuring the preservation of the scientific record. In some instances the scientific society is most likely to assume responsibility. In others, certain publishers will assume it. However, at some point, especially in developing countries, government responsibility for ensuring that provision is made seems appropriate. There was general agreement that authors, scientific publishers, learned societies and governments share some responsibility for financing maintenance of the record. The specific provisions are likely to vary among the countries. There are some strong national interests in maintaining the record of science, but the long term interests are clearly international whether viewed by scientists, users, publishers, societies or governments [Recommendation II.1].

The concept of an electronic archive was based primarily on current experience rather than a more ambitious vision of what scientists and those who benefit from scientific achievements will need from an electronic record. However, the promulgation of standards is the key strategy for stimulating the development of electronic archiving of science. Standards are required for eligibility, for maintenance and for access, along with standards for the structure and content of electronic archives [Recommendation II.2].

Consideration was given to what material should be eligible for the electronic archive and whether legal deposit carries some assurance that intellectual property rights will be respected. There was moderate agreement that only "published" material should be eligible for an electronic archive. However, this view held primarily due to lack of a clear concept of what types of "unpublished" material might be significant additions to the record [Recommendation II.3].

The concern for restricting access to the record established by legal deposit was challenged by the need for scientists to have relatively free access to the record. The scientists also insist that the electronic record be operable to enable them to navigate between their current work and the electronic record quickly and easily [15].

4.3. WORKING GROUP 3: DEVELOPING COUNTRIES

Twenty-six people participated in this working group. Included were academics and publishers from four developing regions of the world – Asia (India and Nepal), Latin America (Brazil and Mexico), the Middle East (Lebanon and Turkey), and Sub-Saharan Africa (Cote d'Ivoire, Malawi, and Uganda). In addition, three European publishers attended the session. There were representatives of several international organisations, universities, and government agencies.

It was recognised initially that developing countries should not be grouped together as a single entity because political conditions and infrastructure vary within regions and also from region to region. In addition, circumstances can diverge considerably within countries. Participants accepted that ôelectronic publishingö should be understood here in the broader sense (comprising production of scientific literature, data banks, indexes and other information relevant to science and technology) and that this definition included electronic communications since e-mail is often the first form of Internet activity to be provided in developing countries.

Issues addressed by this group were as follows:

1. Who is responsible for producing electronic publications in the developing world?
2. How can electronic publishing be used to improve the visibility of developing countries' science?
3. Who is responsible for extending networks in developing countries?
4. What technologies are advisable for electronic publishing in developing countries?
5. Will electronic publications in the long run be less costly (more accessible) for the end- user in these countries than print publications and what strategies can be adopted to this end?
6. What is the role of international cooperation?

The first half of the discussion concentrated on the first three questions because of their interrelation. The responsibility for producing scientific literature (print or electronic) rests in the hands of the local scientific communities and institutions. However, scientific publishing in many developing countries is of a very fragile nature. Scientists often prefer to publish in North American or European journals rather than in journals of their own country or region. Often academic and research institutions in developing countries value these journals more highly, as do many scientists. There is a perception that "local" journals are not as good. As a result changing the medium, from print to online, will not improve the situation of endogenous publishing unless the scientific communities of developing countries re-evaluate and strengthen their own publications. Governments can play a role, as well. In both Mexico and Brazil, for instance, the Councils for Science and Technology have set up committees to establish criteria and evaluate local journals.

A related problem is the fact that scientific research in developing countries is often unknown to the rest of the world and sometimes not even circulated outside of the institution where it is being carried out. It does not appear in the international indexes nor do local indexing and abstracting services fill the gaps. The new technologies should be used to the maximum extent possible to produce appropriate indexes and databases of developing country research information, and also take advantage of specialised international indexes, such as Medline and CAB Abstracts.

Developing countries need to create their own databases and electronic products, and also need to take more control over the technology itself. Successful software development in India, Mexico, and Brazil was cited as an example. Thus, there was consensus among the participants that electronic publishing can ameliorate many of the problems sketched out above-for those publishers who already have the technology at hand, the capacity, and the material to publish, and for those users who are already plugged in.

Governments play different roles in different countries. Everyone agreed that they have major responsibilities for infrastructure strengthening as well as promulgating regulations and tariffs that are favourable to connectivity. Although some governments are more sensitive to the need to implement Internet connectivity, others place Internet at a lower priority. It is up to the local communities, the users and producers of information, to press for an enabling environment. This includes the private sector. Moreover, this "enabling environment" is more than laying lines or installing computers, it also encompasses training end-users how to use the technology to its full extent [Recommendation V.1].

Rather than depending on aid programmes developing countries must take on these responsibilities for themselves. Implementing and using electronic connectivity can be accomplished without huge outlays of funding, provided there is sufficient organisation and planning. Sharing costs among institutions for high-speed links may be cheaper and more efficient, for example, than providing a dial-up link for each person [Recommendation V.2].

Moving on to question 4, participants agreed that no one medium is sufficient of itself. A mix of technologies, print, CD-ROM, e-mail, and interactive Internet, should be employed, depending on the needs of the user, the nature of the product, and local infra-structure conditions. Local publishers and information sources should begin digitising their information as quickly as possible, in whatever form is most appropriate, but taking into consideration the need to observe standards and to ensure convertibility of databases. In addition, it was hoped that electronic publishers in the North would be sensitive to the needs of developing country scientists who cannot access online journals and thus consider production of their digitised information on compact disc as well. It was emphasised that CD-ROM is not a dying technology; it is used in academic libraries in North America and Europe, and is very appropriate to developing countries world-wide. Mexico, Brazil, Zimbabwe, India and South Africa, to name only five countries, are active CD-ROM producers.

With regard to questions 5 and 6, it was urged that UNESCO and other United Nations agencies should continue to provide financial support to encourage further co-operation [Recommendations V.3 & V.4].

4.4. WORKING GROUP 4: IMPLICATIONS FOR TRAINING AND WORK-STYLES

The first matter discussed by Group 4 was the extent of the diversity of information-gathering methods of scientists. Each scientific discipline has its particular methods for gathering and disseminating information, and the needs of scientists will also vary according to other factors such as level of responsibility, place of work or language. A range of basic skills common to most scientists can nevertheless be identified. In particular good authoring skills, adapted to the electronic environment, will be required for rapid and efficient dissemination of information through electronic publishing techniques. In addition, supporting staff such as technicians, network specialists and managers should receive special training so that they can contribute to the development of scientific electronic publishing.

The skills for accessing and effectively using electronic publications should be seen as part of a broader required competency in information retrieval techniques and computer and network literacy. Additional skills are needed for preparation of electronic articles and for the organisation of mechanisms for their dissemination. All scientists should receive training in these areas. Training in information resources and library use should be a part of all scientific education, if possible as early as the undergraduate level, since students can learn these skills more quickly and naturally than scientists who have already acquired work habits. This training should stress characterisation of and access to scientific information including theory and practice of retrieval techniques, as well as general authorship skills including proper structuring of information for publication. Basic training should also aim at providing an understanding of the principles, main activities and tools (web servers, compression techniques, etc.) required for the electronic production and dissemination of information, but need not include detailed practice in this area.

All institutions of higher learning in science and all major employers of scientists should ensure that their students (and employees) receive appropriate training in use of information resources. Such training should take place so far as possible in a discipline or employment specific context, and should take full account of existing training opportunities including commercial offers. Information training activity should be co-ordinated at the level of each organisation, and libraries are often seen as the most appropriate institutional setting for this function [Recommendation IV.1].

In considering what additional factors can facilitate access to and dissemination of electronic publications, it was agreed that adequate and reasonably priced network access is essential for scientific work today. Although cheap, high capacity, data communication provision is predicted for the future, there is a concern that available band-width will be insufficient in the medium term to take care of scientific information needs. particularly in the context of competition from non-scientific network applications. One, technically feasible, solution would be for national authorities to ensure priority network access for bona fide scientific usage. The introduction of a common system of meta-information to identify scientific information resources would not only facilitate access and retrieval, but would also reduce demands on network and server capacity caused by repetitive, "brute force" inventorying of electronic information resources. Cooperative research by the scientific, library and information communities will be essential in finding appropriate cooperative solutions to the organisation of electronic scientific information. If current trends continue, it may also become necessary for scientific communities to consider limiting access to their servers by general network probing programs [Recommendation IV.2].

The Group then considered what new developments are likely and whether they would require training. Clearly, further developments in electronic publishing will increasingly influence the work of scientists. The field of electronic information services, and of electronic publishing in particular is evolving so rapidly that it is probably not fruitful to try to anticipate specific developments. The use of specific tools and techniques can be taught as required, or even self- learned, once scientists have acquired a basic information culture.

It was also considered whether libraries provide dedicated technical support for aspects of electronic information. The information resources required in and for science are becoming ever more numerous and complex, with the result that intermediaries will probably always be needed even as working scientists take more responsibility for gathering and dissemination of information. Other types of technical support are also needed to ensure effective access to and use of information – for example advice on the development and interconnection of computer networks. Operational and methodological support in the handling of electronic information should be organised for individual scientists and scientific institutions. Libraries, which are increasingly assuming gateway and publishing functions in addition to their traditional roles in classifying, storing and disseminating documentation should help fulfil this role. This implies a new type of training for information professionals to be able to perform this function [Recommendation IV.3].

As electronic publishing develops various scientific communities will increasingly be called upon to take more responsibility in the information transfer process, including some functions that have been traditionally the responsibility of print-on-paper publishers. The international scientific community should facilitate exchange of experience between scientists and scientific disciplines concerning electronic publishing applications, with a view to identifying best practice and encouraging cooperative research and development. This is currently under investigation by ICSU Press and it is planned to establish an information network to refer users to information resources, tools and standards relating to electronic publishing in science. Top level access to this information will be incorporated into the ICSU World-wide Web site [Recommendation IV.4].

4.5. WORKING GROUP 5: PEER REVIEW

It was generally accepted that peer review, largely on the basis of the print-on-paper model, is essential although the point was freely made that full advantage should be taken of the facilities for the presentation of sound, film and computer modelling [Recommendation I.1].

Techniques for establishing priority of publication were touched upon, and it seems that the requirements differ from science to science. In high-energy physics, priority may be established by a comment in a public conference, but in other sciences, where interpersonal competition may be greater, stricter criteria prevail.

It was accepted that theories and the interpretation of results can change, but the raw data should be reliable and unquestioned. The tracking can be done by adding tags or links to subsequent revisions. It was even suggested that the changes made during revision of a manuscript should be flagged and made available to the public, noting that a lot of useful material was generated in the path between referee and author and that it was a pity to lose it. This applies equally in the case of pre-print bulletin boards (or e-print archives) which provide a means for discussion of new topical research. Originally established by Ginsparg [6] for high-energy physics theory, their use has been extended to other fields such as molecular biology, particle physics and condensed matter physics. They are particularly important for the development of ideas at the frontier of rapidly developing sciences and are much consulted by researchers. An audit trail was included in the CIF crystallography data base that tracked any changes made to the information stored and this had proved extremely valuable. Some means for electronically date-stamping electronic manuscripts was needed, although it would be difficult to do this in a way that could not be subsequently altered. The quality control and validation of some databases also require attention, for it is desirable to check the validity of the material deposited within it, and to prevent subsequent manipulation of the information. It was clear that it is harder to avoid the doctoring or amendment of digital material than with printed documents [Recommendation I.2].

Quality control was regarded as crucial. This covered the need for scientific validation, good copy editing and the deposition of the definitive version of a paper. Here there could be a bonus in the electronic world by tracking of the way in which thoughts were expressed and concepts established [Recommendation I.3].

There was discussion of the definition of "publication" and it was agreed that presentation of information or observation on web home pages should be regarded as publication. The value of moderated discussions was also considered. Should they be preserved, reviewed and edited, and if so, by whom? How could the possible legal complications for the publishers (who would be accorded responsibility for libellous statements) be avoided?

Finally, the quality marking of papers in an electronic database for the benefit of those unfamiliar with the subject was considered. One way could be for learned societies to attach their logo to the journal title as a mark of authority. Another could be to use an R in a circle, rather like the copyright symbol, to indicate that the paper has been properly refereed [Recommendation I.4].

5. Reports of the Conference

The full text of each of the papers from the Conference is available on the World-wide Web in the ICSU Home Pages [7] and has been published in hard-copy [8]. There have been eight summaries so far published since the Conference and these have all been prepared from different viewpoints. The article by Declan Butler [9] was the first to appear the week after the conference and highlighted the discussion on Peer Review. Dennis Shaw [3] emphasized the recommendations which were published alongside his summary of the papers presented in each session. Lex Lefebvre of STM International [10] emphasized the importance of the various recommendations addressed to publishers in general whereas Hans Sens [11] concentrated on the way electronic publishing had already transformed the process of publishing and outlined the plans of physicists to re-engineer the whole scene by the construction of a seamless electronic web throughout the chain of processes leading to the final appearance of the definitive work. David Brown [12] summarized all the contributions in the order they were presented. AnaMaria Cetto [13] summarized the issues facing scientists in developing countries and, finally, the report of the Conference Chairman Roger Elliott [14] identified those issues which are of concern to the four major constituencies; scientists, librarians, information brokers and publishers, and drew attention to the far-reaching nature of the recommendations.

6. Widening the consensus to cover the international scientific community

As indicated in Section 3.1, the recommendations were reported to the delegates at the 25th ICSU General Assembly in Washington DC and arrangements were made for a special open session for discussion. To set the scene, Scott Lubeck, Director of the National Academy Press explained the development of a Website publicizing the activities of the NAP which had been stimulated by his participation in the Paris Conference.

Roger Elliott continued with a brief summary of the Conference recommendations in which reference was made particularly to the section which dealt with peer review and codes of practice. It was recognized that peer review would differ considerably if conducted electronically since a much wider audience could have access to a text submitted for publication before it was formally accepted. A code of practice was necessary to safeguard an authorÆs intellectual property against plagiarism. It was also necessary to introduce coding to identify the status of a document [Recommendation I.2].

Dennis Shaw explained the need for the creation of an electronic archive the contents of which would need to be kept under review (as were other text archives) if it were to have a lasting value. It was recognized that Learned Societies could play an important role in the establishment and maintenance of archive material but funding was a matter of concern since it was unlikely to be commercially viable. This problem was particularly severe in those disciplines such as earth sciences and astronomy where modern techniques of measurement generated vast amounts of data. Reference was made of the plan for Elsevier to create a databank of all back issues of their science journals. This could be a major resource and its mention led to a discussion how to change the current practice of scientists in giving unrestricted transfer of the copyright of their publications to the publisher: ICSU was urged to address this problem. It was felt that the use of copyright to restrain the free dissemination of abstracts was particularly unfortunate. The matter has since been addressed by the World Intellectual Property Organization at a three-week diplomatic conference in Geneva in November 1996. One important agreement was that the existing criteria relating to copyright exceptions, in particular fair dealing for research or private study, in respect of printed works are to be retained for digitally stored works. A decision on copyright protection of databases was deferred until the Spring 1997 [15].

Kai-Inge Hillerud in presenting the section on financial considerations asked how funding agencies could be persuaded to include in grants the costs of publication and accessing the results of research. It was agreed that this clearly needed action by the international scientific community. The costs and benefits of electronic publication were to be the subject of a detailed study which would involve all sectors of the information creation and publishing chain. The need for such a study was accepted by the audience [Recommendation III.3].

Howard Moore referred to the importance of training scientists in best methods for depositing and accessing information. It was agreed that advice on current practices should be collated and made available through the ICSU Press Home Pages. The Scientific Unions should be consulted to establish what was already available in each discipline. The needs of scientists in developing countries were very much the concern of UNESCO which was planning to give support to pilot projects in this field.

Ana Maria Cetto presented the special needs of developing countries and referred to ways in which the conference initiatives were being followed up. A summary of possible follow up activities has been prepared by Irving

Lerch [16].

It was concluded that the recommendations were clearly acceptable to the international scientific community and there was no dissension expressed in the general thrust of the conclusions from the conference. ICSU Press would accept responsibility for progressing and monitoring follow-up action on these recommendations. This was strengthened later during the day when the General Assembly included among the Resolutions the following:

> The 25th General Assembly of ICSU, recalling Resolution II.9 adopted by the 24th General Assembly; thanks ICSU Press for having organized an Expert Conference on Electronic Publishing in Science, in February 1996; notes the Recommendations in the report of the Conference and invites ICSU Press to undertake appropriate follow-up activities, and to continue to monitor progress in this field, in particular with regard to developing countries, and to report to the Executive Board.

7. Subsequent action

To undertake the tasks laid on it by the General Assembly, the Standing Committee of ICSU Press has been strengthened under the chairmanship of Roger Elliott. It is proposed that ICSU Press should act as a focus for a programme of follow-up activities which will also involve a number of other international bodies. Discussions are already in progress with the International Council of Scientific and Technical Information (ICSTI) and the International Federation of Library Associations and Institutions (IFLA), the American Association for the Advancement of Science (AAAS) and the Advisory Council for Scientific Communication of UNESCO, and there are several other bodies with a similar concern. The three general issues of most concern to ICSU, deriving from the Conference, relate to the needs for a code of practice for electronic publication in science [Recommendation I.1-4], the establishment of principles and guidelines for archiving [Recommendation II.2] and the study of the economic costs and benefits of electronic publishing [Recommendation III.3]. ICSTI and IFLA have shown a particular interest in the archiving issues while UNESCO is particularly concerned with the impact of electronic publishing on developing countries and the Recommendations V.1-4.

Within the ICSU family itself, the Chairman of ICSU Press has instituted an enquiry to establish the use and experience of members with electronic publishing. In order to facilitate the dissemination of information about these issues it has been decided to establish an ICSU Press Website that will be accessible directly and also via the ICSU Home Pages [Recommendation I.4]. Discussions have also begun on how best to adapt the ICSU Home Pages to provide an information network [Recommendation IV.4]. Dennis Shaw has been appointed to manage this project

ICSU Press will seek funding to support a residential workshop for a technical study of the economics, costs and benefits of electronic publishing [Recommendation III.3].

8. Membership of Committees

8.1. PROGRAMME COMMITTEE & PROCEEDINGS EDITORIAL BOARD

8.1.1. *Full Members*
- Dennis Shaw (*ex officio*) **Chairman**
- Bryan Coles (UK, Director of Taylor & Francis)
- Roger Elliott – (Conference Chairman – *ex officio*)
- Ekkehard Fluck (Germany – Director of Gmelin [CODATA])
- L J Haravu (India, Information Manager of ICRISAT)
- Tarja Koskinen-Olsson (Finland, Chairman of IFFRO)
- Christian Lupovici (France, Assistant Director of INIST)
- Vitaly Nechitailenko (Russia, IUG)
- Marthe Orfus (ICSTI), **Secretary**
- John Rose (UNESCO)
- Anthony Watkinson (UK, Director of Chapman & Hall)

8.1.2. *Corresponding members*
- Graham Cornish (UK, British Library Copyright Officer)
- Thomas Dreier (Germany, Max Planck Institute for Patent Law and Intellectual Property)
- Howard L Funk (USA, Vice-President of IFIP)
- Arnoud de Kemp (Germany, Director of Springer, Chairman of STM Innovations Committee)
- Ivan Klimes (UK, Director of Rapid Communications of Oxford)
- Lisbeth Levey (USA, Director of AAAS Sub-Saharan Programme)
- Brian McMahon (UK, Project Leader IUCr)
- Geert Noorman (Netherlands, Director of Elsevier Science BV)

8.2. INTERNATIONAL ADVISORY COMMITTEE

- Adnan Badran (UNESCO Assistant Director for Science)
- Georges Cosmides (USA, NIH Toxicology Programme, NLM)
- Mike Dacombe (UK, Executive Secretary of IUCr)
- Mihály Ficsor (Assistant Director General of WIPO)

- Eugene Garfield (USA, Honorary President of ISI)
- Lex Lefebvre (Netherlands, Secretary of STM)
- Edward N Maslen (University of Western Australia)
- Franco Mastroddi (DG XIII – EEC)
- Simon Mitton (UK, Science Director, Cambridge University Press)
- Kurt Pawlik (Germany, President of IUPsyS)
- Walter G Peter III (USA, Deputy Executive Director of ASM)
- David Russon (UK, President of ICSTI)
- Kent A Smith (USA, Past President of ICSTI)
- Juan Voutssas (Mexico, Director of UNAM Information Center for Science and Humanities)
- Mark Wahlqvist (Australia, Monash University [IUNS]).

8.3. ICSU PRESS STANDING COMMITTEE

- Chairman: Roger Elliott (United Kingdom)
- Liu Zhadong (China)
- H. Spekreijse (Netherlands)
- Scott Lubeck (NAS-USA)
- V. A. Nechitailenko (Russia)
- K-I. Hillerud (Sweden – ICSU Legal Adviser)
- H.J. Moore (UK UNESCO representative)

References

1. Shaw, D.F. (Ed.) 1993, The bibliographic control and protection of intellectual property of digitally stored texts in the scientific domain: report of the ICSU/UNESCO meeting of experts held at UNESCO Headquarters, Paris, 7-8 June 1993. ICSU, Paris (ISBN: 0-930357-30-2).
2. URL: http://www.lmcp.jussieu.fr/icsu/Information/Proc_0296/
   or http://www.grainger/uiuc/edu/icsu/confchmn.htm
   or http://www.thomson.com:8866/ICSU/
   or http://www.eos.wdcb.rssi.ru./eps/
3. Shaw, D. 1996, *Vistas in Astron.* **40**, 369-380
4. Campbell, R. & Russon, D. 1996, *LOGOS* **7**, 178-185
5. Lustig, H. 1997, Electronic publishing: Economic issues in a time of transition, this volume
6. Ginsparg, P. URL: http://xxx.lanl.gov/blurb/
7. ICSU Home Page: http://www.lmcp.jussieu.fr/icsu/
8. Shaw, D.F & Moore H.J. (Eds.) 1996, Electronic publishing in science, ICSU Press & UNESCO, 198 pp. (ISBN 0-930357-37-X)
9. Butler, D. 1996, Peer review æstil essentialÆ, say researchers *Nature* **379**, 758
10. STM 1996, ICSU Press – Unesco Expert conference on electronic publishing in science, held in Paris at Unesco, 19-23 February 1996, *STM Newsletter* **99**, 8-10
11. Sens, J. 1996, Electronic publishing in science, *Europhysics News* **27**, 68-69
12. Brown, D.J. 1996, Electronic publishing in science, *icsti forum* **22**, 5-7

13. Cetto, A.M. 1996, Science International **61**, 29
14. Elliott, R. 1996, Chairman's report, in *Electronic Publishing in Science*, Eds. D. Shaw & H. Moore, ICSU Press & UNESCO, Paris, pp. 13-15
15. WIPO Conference 1996, Draft Recommendation CRNR/DC/88 – see also *Nature* **384**, 293
16. Lerch, I. 1996, private communication

Appendix: Recommendations from the Conference

The Conference Recommendations have been published in Science International, Newsletter No. 61, 1-3. (ISSN 1011-6257 May 1996). They are also available online at the following URLs:

http://www.lmcp.jussieu.fr/ICSU/
　　　　　　Information/Proc_0296/recommendations.html
http://www.surya.uiuc.edu/icsu/recommendations.html
http:/eos.wdcb.rssi.ru/eps/.

These recommendations focus attention on the actions which can and need to be taken during the remaining years leading up to the beginning of the 21st century to ensure that the maximum benefit is derived for all concerned with the progress of scientific research by the application of information technology to the distribution of scientific information. Action following the adoption of these recommendations by ICSU in October 1996 will be decided by members of the ICSU family in consultation with UNESCO where appropriate.

The Conference of Experts on Electronic Publishing in Science convened by ICSU and UNESCO in February 1996 discussed the broad range of problems and opportunities presented by the new technologies. It approved a number of recommendations, roughly grouped below under five headings, which are directed specifically to ICSU and UNESCO and through them to the scientific community in general, the learned societies and national academies and others involved in scientific information provision such as publishers and librarians, and to national governments.

I – Peer review and codes of practice

1. The Conference overwhelmingly recommends that strict peer review should be applied to all scientific material submitted for publication in electronic journals.
2. The peer review system should yield definitive authenticated and dated versions of papers for publication, although the subsequent attachment of tags indicating later developments or revisions of papers should not be precluded.

3. Attention should be given to the definition of acceptable ways of establishing priority of publication, for these appear to differ between scientific disciplines.

4. ICSU and UNESCO are encouraged to organize the most appropriate type of forum involving scientific societies in order to formulate codes of ethics and of conduct for electronic publication which would spell out the reciprocal obligations of the scientist and the community on such matters as peer review, citation, integrity and authentication of material and archiving.

II – Electronic archives

1. Given the traditional role of publication in science in providing an archive, it is timely and urgent that attention be given to ensuring the archiving of science in electronic formats. The malleable nature of digitized information requires the establishment of verifiable electronic archives including databases.

2. The Conference recommends that scientific societies, publishers and librarians come together to establish principles and guidelines for electronic archives covering, but not limited to: maintenance, content, structure, finance, eligibility, accessibility and compatibility. Consultation with the International Organization for Standardization (ISO) should be ensured concerning the use and development of appropriate international standards.

3. It is further recommended that a registry be created for electronic archives in scientific fields, together with the establishment of principles and guidelines for its operation.

III – Financial considerations

1. The Conference recommends that funding agencies regard the costs, both of the publication of research results and of access to required information, as an essential component of research funding.

2. The Conference recognizes that the availability of electronic information in a searchable form is potentially of great advantage to the world scientific community for the efficient conduct of research and education. Funds should be directed so as to allow full use to be made of the potential and development of this source of information.

3. A technical study of the costs and benefits of electronic publication should be carried out by an international committee established by ICSU in coordination with ICSU members and associates, and involv-

ing representatives of the library and scientific, technical and medical publishing communities.

IV – The scientist's working environment

1. Although each scientific discipline has its particular information gathering and dissemination methods, a set of basic skills needed by scientists can nevertheless be identified. All scientists should receive training in information resources and library use and in good authoring skills, adapted to the electronic environment, if possible as early as the undergraduate level. ICSU and UNESCO should assist those developing countries which do not have the resources and expertise to organize this training.

2. The Conference unanimously endorses the view that adequate and reasonably priced network access is essential for scientific work and scientific education. National authorities should ensure that appropriate infrastructure be established for this purpose and that scientific data traffic receives appropriate priority. Scientists in all countries should have good access to computer communications to participate in information exchange needed for their work, and ICSU and UNESCO should promote such access with all available means.

3. In keeping with the increasing role of the scientists in the electronic publishing process, the international scientific community should facilitate exchange of experience between scientists and scientific disciplines in this field with a view to identifying best practice and encouraging cooperative training, research and development. This work should be facilitated by the international committee proposed in Recommendation III.3 above.

4. As a useful first step, an information network should be established to refer users to information resources, tools and standards relating to electronic publishing in science. Direct, immediate access to this information should be incorporated into the ICSU World Wide Web site and facilities should be provided for discussion groups and for E-mail access for those users without WWW capacity.

V – Developing countries

1. The scientific community in developing countries is becoming increasingly involved in all stages of electronic publishing, including locally published on-line journals, CD-ROM technology and appropriate databases. These scientists should assume greater control of the technolo-

gies involved. UNESCO, ICSU and the international community should take this into account in planning and implementing cooperative projects.

2. The scientific community in developing countries should become involved as a partner in the development of methodologies, tools and standards relating to electronic publishing and archiving so that these can be adapted to its needs. In particular, regional cooperation should be encouraged and language-independent systems adopted where possible.

3. UNESCO should be encouraged to support one or two developing country pilot projects in the area of electronic publishing.

4. The costs of infrastructure and information provision are particularly heavy for developing countries. This problem is ultimately the responsibility of the governments concerned. The assistance of UNESCO and other UN agencies in promoting solutions should continue, while better linkages to the local private sector, regional cooperation relying on local expertise, and competitive and flexible regulatory regimes, should all be fostered.

ELECTRONIC PUBLISHING:

ECONOMIC ISSUES IN A TIME OF TRANSITION

HARRY LUSTIG
American Physical Society
304 Chula Vista Street
Santa Fe NM 87501, USA[†]
lustig@aps.org

1. INTRODUCTION

This chapter is an annotated case study of the cost, to the American Physical Society, of publishing its scientific journals: The *Physical Review, Physical Review Letters*, and *Reviews of Modern Physics*. It also reports on the revenues available to cover these costs and on the resulting prices charged to subscribers.

A number of studies, in physics and in other fields, have shown that the costs of journals to subscribers vary greatly from publisher to publisher, whether measured as the price per article, per word, or per character. The differences are significantly exacerbated when the unit price is divided further by the "impact" (the number of citations per year per article) or the "usage" (the number of times per year the journal is consulted in the library). These variations can be explained by several factors, including the very different numbers of subscribers over which the large fixed costs of journals or articles are spread, and different policies regarding profits.

We have chosen the American Physical Society (APS) for a case study of the economics of publishing in the era of transition to electronic dissemination, in the first instance because the author has had access to the needed data. In contrast it is next to impossible to obtain economic data from most other publishers. However, another reason is that APS publishes a major share of the world's physics literature and that its journals are, unarguably, among the most important and prestigious. On the basis of our acquain-

[†]Home address.

Astrophysics and Space Science **247**: 117–132, 1997.
© 1997 *Kluwer Academic Publishers.*

tance with other U.S. non-profit scientific societies, we conclude that APS' motivations and operations are often, but not always typical. The goals and policies of commercial publishers are often significantly different. Throughout this paper we will, wherever possible, take note of both similarities and differences with other physics publishing operations.

Many of the data in this study are for the print journals. A practical reason for this is that, although APS has embarked on a program of having all of its journals on line (in addition to in the print format) before the end of 1997, relatively few on-line (or CD-ROM) subscriptions have so far been effected. This is particularly true for libraries. Related to this situation is the fact that we have not yet settled on a clear, steady state pricing model for electronic distribution. For the moment the revenues from and even the costs of electronic distribution are only a perturbation on the economics of the print journals. Well over two-thirds of the costs of publishing journals are attributable to editing and composing activities – which, unless these operations are fundamentally altered by expectations, problems, and opportunities that are, in part, engendered by the electronic revolution – will not be much affected by changing from print to on-line distribution. We consider these possible "sociological" changes, together with the prospects for generating revenues, in the latter part of this paper.

2. THE ECONOMIC ROLE OF JOURNALS

Scientific publishing, in the age of print journals, has for many years and for many publishers been a profitable enterprise. While privately held commercial publishing companies rarely open their books, reports of large profits have found their way into the popular press. Thus, for example, an article in the December 18, 1995 issue of *Forbes* magazine (whose subject is the threat to the publishers of academic journals from electronic publishing) it was reported that Reed Elsevier had total annual revenues estimated at $5.5 billion. Its academic journals had revenues of $600 million on which they probably earned $225 million, for a pretax margin of nearly 40%.

Not-for-profit scientific societies, particularly in the United States and in the UK, also often realize substantial surpluses from their publishing operations. They use them to support their non-income producing "do-good" activities, such as education, outreach, international programs, and public information, and in some cases, to build up "reserve fund" endowments. Net returns of 30% and more have not been uncommon.

The American Physical Society has consistently budgeted its publishing operations to realize either no or only a modest surplus (10% or less). The Society, which is democratically governed by its members and responsive to a community which includes authors, readers, university and laboratory administrators and libraries, makes its journals available to its members at the "last copy cost", e.g. $145 per year for 52 issues and some 10,000 pages of *Physical Review Letters* for print, and first $75 and soon $25 for on-line access. While prices for libraries are considerably higher than for members, these prices, on a per page or per character basis, are typically one tenth of those charged by commercial publishers. If the numbers of citations are factored in, the cost-effectiveness becomes even greater. The non-profit nature of the APS and self-imposed restraints are not the only reasons for this: the relatively large circulation of the journals, which permits spreading the fixed cost over a larger number of library subscriptions, also plays a significant role.

Even so the surpluses from the publications have been used by the Society to make up for unavoidable deficits in its public affairs programs. As described in the October 1996 issue of *APS News*, the Society has now decided that it must and will wean its other operations away from their dependence on surpluses from the publications. It must do so because first the surpluses and ultimately the fiscal health of the journals itself are in question. It can do so, provided it limits further growth in its public affairs programs, because the Society has over the years built up a reserve fund whose investment returns, in good years, can pay for a substantial share of these programs' expenses. The rest will have to come from contributions and grants.

3. REVENUES

There are two sets of reasons for the fiscal threat to APS' and other publishers' journal operations; both lie in the area of revenues. One is endemic, the other is brought about by the pending transition to electronic publishing.

To understand the reasons for and nature of the endemic threat, we need only to look at the contributions of various publishing revenues and expenses (Fig. 1a and b). Fig. 1a shows, for fiscal year 1995, the sources of income for APS' journals. The predominant source was library subscriptions, which accounted for 74.7% of the revenues. In spite of steady decreases in the number of library subscriptions, this *percentage* (as well as the dollar volume) has been steadily rising. The number of library subscribers to APS' journals – where non-US subscriptions now outnumber domestic ones – has been declining steadily for several decades, at an average 3% per year. The reasons are the rising prices of journals (APS' as well as others); the appearance of new journals; and severe constraints in library budgets.

We have not been able to discern a correlation between the magnitude of the decline in subscriptions and the increase in prices in a particular year,

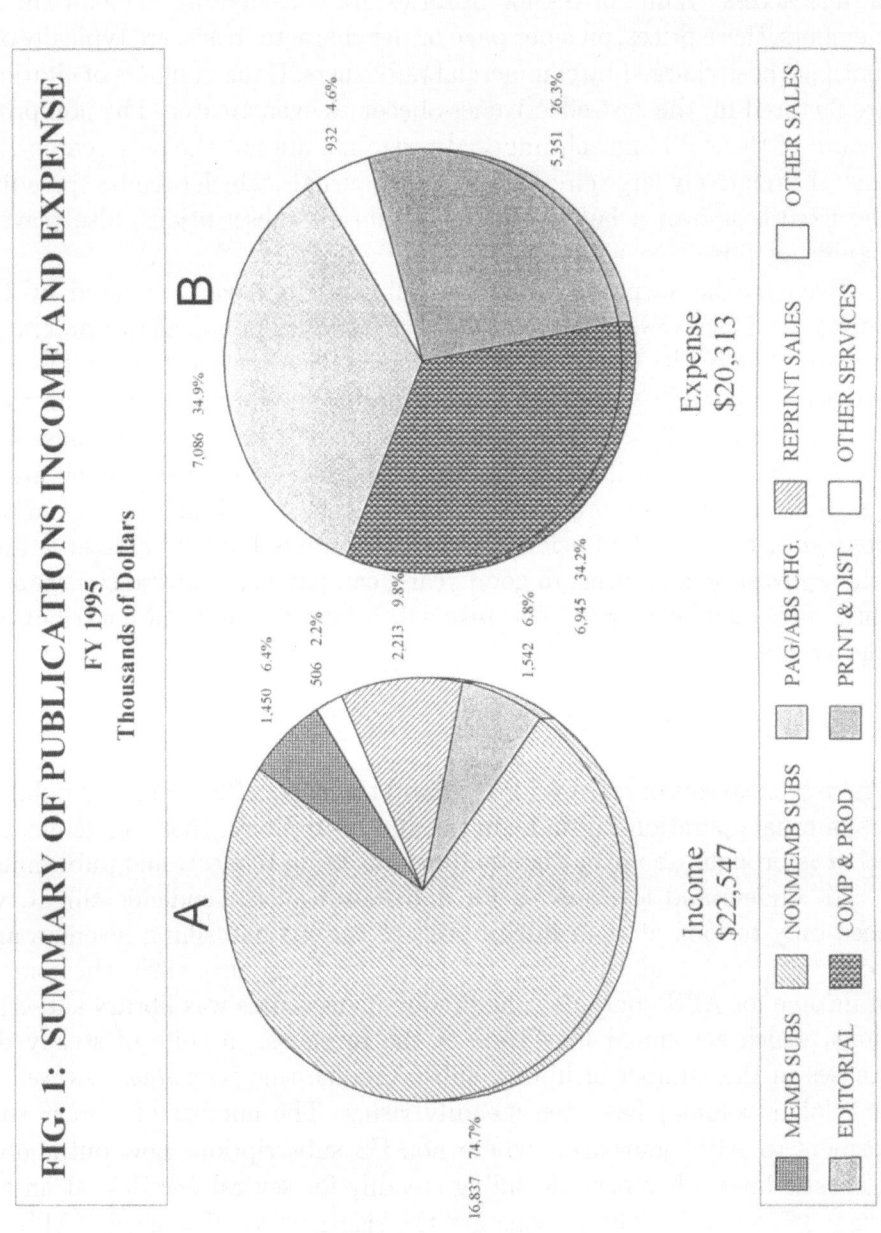

FIG. 1: SUMMARY OF PUBLICATIONS INCOME AND EXPENSE
FY 1995
Thousands of Dollars

A — Income $22,547

B — Expense $20,313

MEMB SUBS   NONMEMB SUBS   PAG/ABS CHG.   REPRINT SALES   OTHER SALES

EDITORIAL   COMP & PROD   PRINT & DIST.   OTHER SERVICES

for APS journals. The price to libraries of APS journals, currently averaging about 12 cents per page, is, comparatively, very low. However APS journals are very large (and have been growing) and therefore their costs, in absolute terms, constitute significant items in library budgets. Some librarians have recently begun to assess cost-effectiveness as "subscription dollar per use", with the denominator measured by the number of times volumes are left on library tables for reshelving. APS' journals have proved themselves as cost-effective by these studies as by their price per character and citation. APS' losses of subscriptions appear to be smaller than those of many other physics publishers. (Publishers rarely release these numbers.) While much of the loss has occurred in institutions that subscribed to multiple copies of a journal, there have also been non-renewals from smaller college and industrial libraries with single subscriptions, both in the US and abroad. And while it is assumed that there is an irreducible number of institutions that must hold on to their single subscriptions, and it appears that cancellations may be slowing down, no one knows when the decline will stop. (In 1995 losses were only 1.9%, but in 1996 they appear to be back to 3%). At the present time there are, on the average, about 1700 library subscriptions to a section of the *Physical Review*, 2300 to *Physical Review Letters* and 2000 to *Reviews of Modern Physics*. We believe that these numbers are generally higher and in many cases substantially higher than those for other physics journals.

Member subscriptions, which are an insignificant source of income for the APS because of their extremely low price based on the "last copy" cost, have nevertheless been declining even faster than library subscriptions. Prices charged to members are so low, because journals are considered a benefit of membership. The much lower prices for individuals, compared to those for libraries, can also be justified because these copies are for the personal use of one person, rather than for many people in an institution. Here not only price, but also shelf space, the small percentage of articles in a journal that any one individual would want to read, and the availability of copies in the library have been significant factors. The seven sections of the Physical Review now average about 500 member subscribers each. *Physical Review Letters* and *Reviews of Modern Physics* are more appealing: they now have about 3,000 and 2,700 member subscribers respectively. Among members, the ready availability, convenience and power of electronic on-line versions of the journals and their even lower price (figured on a "last copy" cost basis) may well bring about an increase in subscribers.

Many other societies do not grant their members as large a discount as APS does. Commercial publishers, who, of course, have no members to whom they are responsible or whom they need to cosset, may nevertheless also distinguish in their pricing between individuals and institutions. When they do, their individual rates are generally not as much reduced from their library rates as those of membership societies. For these publishers individual subscriptions may therefore constitute a more significant source of income.

One of the potential new sources of revenue for research journals is "document delivery" for which a library (or an individual) orders a copy of a single article from a document delivery service or directly from a publisher. This is done when a library has decided not to subscribe to the journal in which the article has appeared and has exhausted its access to free sources under the "fair use" copyright rules. Several universities have recently reported canceling large numbers of journals from one or more commercial publishers and resorting to document delivery of articles in them, when needed, at substantial savings. So far APS' revenues from document delivery have been minuscule. Librarians tell us that they will remain so because our journals are part of "core collections" and it is cheaper and more convenient for the library to subscribe to the journal than to order (many) copies of (many) articles. The situation may however change in an era of electronic distribution.

This brings us to what *was* once a very significant source of income, authors' page charges. In the 1960's and 70's, they contributed over 50% to the revenues of the journals. Now the income is down to less than 10% and fading fast. The reason is not that there is anything ethically or conceptually wrong with page charges. The refereeing of the results of one's research and their communication to other scientists and to the public are an indispensable part of the research process and should be paid for by the sponsor of the research along with the other costs. At about $50 per page (the average APS charge in 1995) they are a trivial component of many research budgets. Because they are voluntary, researchers who do not have external or institutional support, or who have other reasons not to pay them, do not need to do so. There is no discrimination and there are no sanctions. Page charges have the attractive feature of scaling with growth in the number of articles published.

Page charges are collected by a number of US physics, astronomy, geophysics and other journals. In Europe they are looked at askance and commercial publishers generally do not have them in their repertoire, which goes to make their subscription prices even higher. The European Physical Society, regrettably, used to publish an "honors list" of physics journals that did not request page charges. We are pleased that this practice has stopped.

In spite of the virtue of page charges, the American Physical Society is in the process of all but phasing them out. The reason is that because they are truly voluntary; because at a time of tight budgets they have become significant for some researchers, particularly theorists; and because so many of the authors reside in areas of the world where there is either no tradition or no possibility of paying page charges, the honoring rate is steadily declining. Furthermore some researchers, rather than not honoring the page charges, have been driven, or thought they have been driven, to submit their papers to commercial journals that have no page charges. While this has saved the authors or their sponsors some money, it has raised the cost of journal collections for libraries. Because attitudes and behaviors have been different among the different segments of the APS research community, page charges had become a divisive force, which had to be contained.

The phase-down is being accomplished through the device of instituting differential page charges for electronically ("compuscript") and non-electronically submitted manuscripts: lower or zero for compuscripts, higher for conventional manuscripts. The motivation for encouraging the submission in compuscript form is that this will eventually make the cost of electronic production and distribution cheaper. As compuscript submissions become the dominant mode, page charge income will all but disappear. As we will see below, and as some previous critics of page charges are beginning to realize, giving up on page charge income may well be exactly the wrong thing to do in an era of on-line publishing.

The decline in the number of subscriptions and in other income, together with the increase in costs brought about principally by growth in the number of articles and pages but also by unit cost inflation, has forced the American Physical Society steadily to increase its subscription prices. For some years, the increases had averaged 15%. While this is considerably lower than the increases of many other publishers and the prices of APS' journals remain *comparatively* as low as ever, the Society has now embarked on a policy of reducing the price increases through an enforced containment of costs. Average price increases will be held to less than 10% per year. In fact in 1997 they are 9% and for 1998 they are likely to be held to 7%. The most important policy change to enable us to do this has been to limit the growth of the journals.

4. EXPENSES

In fiscal year 1995, it cost APS over 20 Million dollars to accept or reject manuscripts, produce and distribute its journals (Fig. 1b). This is the "fully loaded" cost: it includes, in addition to direct costs, the prorated expense of all technical, administrative and financial services. Also included, under "other services" are the costs of billing for and collecting page charges and reprints and the printing and distribution of the reprints.

The costs that are directly associated with the journals can be readily divided into three categories: editorial (at 34.9% of total costs), composition and production (34.2%), and printing and distribution (26.3%). The production and distribution figures reflect the costs of the print journals only. In our budgets we are committed to integrating the electronic production and distribution costs with those for the print journals: we want to get away from seeing electronic publishing as an add-on activity. However some of the costs of on-line distribution are not yet well known. For the purpose of this presentation, we will therefore deal with the costs of on-line and CD-ROM publishing in a separate section.

4.1. EDITORIAL COSTS

Some readers may be surprised at the high costs of receiving, refereeing, accepting or rejecting manuscripts and preparing them for composition. These costs amounts to about $350 per manuscript received, $500 per manuscript published, or $70 per page published. (These numbers represent alternative ways of stating the cost: they should not be added.) While APS' editorial costs may be somewhat higher than those of some other publishers, there is evidence that they are not out of line. APS pays neither its authors (except for nominal honoraria to the authors of review articles in *Reviews of Modern Physics*) nor its referees. In 1995 over 20,000 manuscripts were received, involving more authors than that number. APS' referee database includes 26,114 names. Of this number, 12,655 reviewed at least one manuscript in 1996 and many were used more than once. But while authors and referees come free, mediating between and otherwise interacting with them, and vetting and preparing the manuscripts for publication decidedly does not.

Manuscripts must be received, logged in and classified. Referees must be selected and engaged, and the timely submission of their reports must be assured. When reports do come back, additional refereeing is sometimes decided on, either before or after the authors have seen the referees' comments. Even before an editor makes a decision whether or not to accept a manuscript for publication, a sometimes protracted interchange between an author and the referee may take place, with the editor acting as middle man. (The identity of the referee is not revealed to the author.) If and when an editor makes a decision to reject a paper, authors frequently appeal. The appeal process may require another cycle of reviews by new referees. The editors of other journals generally accord authors much less opportunity to "negotiate" with referees and many journals have no appeals process whatever. APS, as a democratic society, is beholden to its members and authors and, with a long-held reputation for fairness, cannot run the risk of appearing high-handed.

Another area in which APS' editors have provided more services than those of other journals is in editing manuscripts for publication. This "pre-marking" has included not only cleaning up scientific ambiguities and poor English, but also imposing a measure of uniformity in nomenclature, style and format on the articles, and to a degree, even between journals.

To carry out these responsibilities and provide these services, APS has assembled a staff of over 100 in a central facility at Ridge on Eastern Long Island of New York. (The location is close to the Brookhaven National Laboratory, where APS' editorial operations were carried on until the seventies.) About 25 are physics PhD's and another 25 are scientifically trained persons who are also engaged in editorial work; the rest constitute the support staff. In addition to the full time resident staff, most of the journals also have "remote" part-time senior editors, with small support staffs, at universities, and there are some 30 remote associate editors and about 135 members of editorial boards in various locations throughout the US and abroad.

The maintenance of the large central facility and staff is highly unusual. Most other scientific journals are edited and prepared for production at universities and other centers of research, under the part-time direction of a local scientist-editor and with a small local staff. In the past many of these operations have been generously subsidized by the host institution and some editors have worked without remuneration, for the honor of it.

The centralized-cum remote style of APS' editorial operations tends to increase the cost. One reason is precisely that the editors of the different journals do not work in isolation but strive to maintain consistent if not uniform policies and style across the journals. This produces a desirable result but takes time and effort. Another good outcome of having a central core of fully engaged physicist-editors has been the increasingly fruitful interaction between editors and the leaders of the Society who make the publishing decisions, something which usually does not as readily occur in commercial publishing enterprises or even in some other scientific societies. The editors' advice and participation have been particularly valuable in planning for and launching the transition to on-line publishing.

Short of decentralizing the Ridge operation — something that is not in the cards – how can we lower or at least contain editorial costs? The refereeing process should and can be shortened, principally by limiting the number of appeals. Steps to do that have already been taken. Another measure is to reduce editorial "pre-marking". This too has already been instituted, albeit with regret on the part of some editors and users. The potentially most significant step to contain costs is to halt growth or even to reduce the size of the journals. Unless submissions decline (something which is not likely to happen in the immediate future), the acceptance rate will have to be lowered. Taking this step has been ordered by the APS Council and has begun to be implemented. Because authors may and frequently do submit papers rejected by the *Physical Review* and by *Physical Review Letters* to other journals, where they are accepted and published, this act of self-denial by the APS is not likely to save any money for libraries. But it will help to keep the costs and prices of APS' journals down and their quality up. Greater selectivity in accepting papers for publication will however not save much immediately, even for APS: it has been noted with only slight exaggeration that it costs more to reject an article than to accept and publish it.

4.2. COMPOSITION AND PRODUCTION COSTS

The composition and production of the APS journals also cost about $500 per published manuscript or $70 per published page. This includes the keyboarding of manuscripts, the preparation of illustrations and color figures, and copy editing; in other words all the steps that are needed before the printing presses are set up and run. For electronically submitted compuscripts and for electronic production and distribution, there is tagging and mark-up. APS does not do any composition and production "in house", with the exception of some developmental work in electronic publishing. All sections of the *Physical Review* as well as *Reviews of Modern Physics* are composed and produced for APS by the American Institute of Physics (AIP) for a fixed, negotiated price per page. *Physical Review Letters* is produced for APS by the Beacon Group of Ashland Ohio. The use of AIP, for most of the journals, is cost-effective because of the large number of journals and pages – they include AIP's own publications – handled. APS maintains *Physical Review Letters* at Beacon in order to have an independent check on price and performance.

Together, the editorial and the composition/production expenses constitute the fixed or "first copy" costs of the APS journals. They account for almost 70% of the total cost of the journals. For publishers whose journals are sold to fewer subscribers, the fixed costs are an even higher *percentage* of the total. These costs are incurred before a single copy is printed or electronically distributed. They do not rise if there are more subscribers and they do not decrease if there are fewer. The small number of subscribers to scientific journals largely accounts for the high price of subscriptions; the greater the number of subscribers, particularly library subscribers, the lower the price of a subscription can and should be. Assuming that the fixed costs are to be recovered entirely from library subscriptions – a pretty good assumption with the rapid decline in page charge income – the fixed costs per subscription for *Physical Review B* in 1995 – 37,472 published pages, 1865 subscriptions – were $2693. This calculation takes into account the editorial cost of rejected as well as of accepted papers. *Physical Review B* is the largest journal published by APS, but also has more library subscribers than any other section of the *Physical Review*; the costs per subscription for the other journals scale directly with the number of pages published and inversely with the number of subscribers.

4.3. DISTRIBUTION COSTS

For print journals, distribution expenses include the costs of paper, printing and binding, preparing for mailing, and postage. To these must be added the costs of subscription fulfillment and servicing. APS uses two commercial printing firms and, with the help of AIP, shops around for the best prices for the quality of paper needed to ensure the long life that is required for our archival journals. Two years ago, there was a dramatic increase in the price of paper, that was apparently caused by heavy demand from mail order advertisers. Printing costs consist of a flat charge for setting up and starting the presses plus a charge depending on the number of pages printed. Domestic and overseas surface delivery is entrusted to the US Postal Service, which also increased its prices substantially a year ago. Air freight delivery, which is handled by private companies, adds substantially to the cost of distribution overseas.

Subscription orders, renewals and fulfillment for members are handled by APS' membership department with the participation of AIP. Orders for library subscriptions to the APS journals are handled by AIP, almost entirely with the intermediation of subscription agencies. AIP charges APS $6.10 for the fulfillment of each journal subscription and states that it is losing money on the deal.

This discussion shows that distribution costs have a fixed component that scales with the number of pages distributed and a component that is proportional to the number of subscriptions maintained. Thus there is no single figure of merit for the unit costs of distribution. As a very rough estimate, an average of 1.8 cents per exemplar, per page may be used. Thus distribution expenses add about $675 to the cost of a subscription to *Physical Review B*. About half of this amount constitutes the true "last copy" costs.

5. THE FINANCES OF ELECTRONIC PUBLISHING

There are individuals in this world, scientists and others, who think that electronic on-line publishing is or ought to be free. An example that is cited in defense of this proposition is the very useful and highly regarded physics e-print service established and maintained at the Los Alamos National Laboratory (LANL) by Paul Ginsparg and his associates. It is true that users of this service pay little more than network connection charges. However, the installation and maintenance of the service has been heavily subsidized by LANL and, recently, through large grants from the National Science Foundation. Nevertheless, the costs incurred by the "publishers" for receiving and distributing the articles are significantly lower than those for APS' and other publishers' journals. One crucially important reason is that the LANL e-prints are not refereed or edited.

We agree however that the costs of printing and distributing on-line and CD-ROM journals ought to be lower than those of print journals. This will actually be the case when R&D investments have leveled off and when reliable, standard technologies have been developed and adopted. Unfortunately lower costs do not necessarily guarantee fiscal stability and the flowering of electronic journals. The reason is that in the era of the Internet, a culture of "free" access to on-line information, and a time of uncertainty about copyright, combined with technological means for easy copying, it is not at all clear who will pay for electronic access.

5.1. COSTS

It would be easy to conclude that the fixed costs of publishing – which, as we have shown, constitute, for APS, about 70% of the total - will not be affected by electronic distribution. This is indeed largely true for the *editorial* costs, except that some savings will be realized when manuscripts are seamlessly written, submitted, sent to referees, returned with comments and revised electronically, a prospect which the APS is actively pursuing.

At present there are also no savings in the composition and production costs for electronic journals; on the contrary, there is an additional expense. The reason is that electronic "tagging" still occurs *after* the pages are composed for print production. The additional costs are about $10 per page. If and when the electronic composition occurs "up front" or in lieu of composition for print, and if and when electronic publishing takes place as a single continuous process from the submission of a manuscript to its distribution as a published article, costs will be significantly reduced.

It is in the area of distribution where on-line as well as CD-ROM publishing already realizes significant savings over print. However, this sector, even with APS' relatively large number of print subscriptions, accounts for only 26% of the cost. Unit cost estimates for electronic distribution are very much in flux, but seem to be settling in at about 1.0 cent per exemplar page, instead of the 1.8 cents for print, with large savings in paper and postage being partially offset by considerably higher subscription fulfillment and subscriber service costs.

Table 1 summarizes APS' **add-on** budget for electronic publishing in fiscal year 1997. It includes both operating and research and development expenses.

TABLE 1. APS EP budget in FY 1997

Item	Budgeted Expenses
Journal Information Systems Department	734,000
Physical Review Letters on-line	261,000
Physical Review C on-line	46,000
Physical Review D on-line	184,000
Other journals to go on line	100,000
Archiving of Physical Review and Physical Review Letters 1985-1995 (PROLA)	193,000
Other projects	155,000
Total	1,673,000

5.2. PRICES AND REVENUES

APS began distributing on-line and CD-ROM versions of its journals to members and libraries on July 1, 1995. This is when *Physical Review Letters* went on line. *Physical Review B Rapid Communications* (for members only), *Physical Review C* and *Physical Review D* followed during 1996. The pricing for these subscriptions has been based on the following principles and guidelines.

1. APS will not attempt to recover its R&D expenses over the short run.
2. Except for these R&D expenses, prices ought to reflect actual costs (prorated editorial and actual production and distribution costs for libraries; distribution costs only for members).
3. Electronic subscriptions are to be available independently of print subscriptions, but at least for the time being may also be obtained, by

libraries, for an appropriate extra charge in connection with print subscriptions.

4. Electronic subscriptions will not be given away free, except for short periods of possible unreliability, when they will be accessible free of charge with a print subscription. (This provision contrasts with the practice of some other publishers who have opted to give away "free" electronic access to print subscribers for extended periods. The word "free" is in quotation marks because in fact, print prices have sometimes been raised an additional amount to compensate for the cost of the electronic production and distribution.) APS does not believe in trying to habituate its subscribers to its on-line journals in this way. Other publishers are learning that when the on-line give-aways cease and are replaced by reasonably priced subscriptions, the withdrawal symptoms suffered by subscribers are not strong enough to result in significant numbers of "renewals".

These principles and policies have resulted in the prices and subscriptions for *Physical Review Letters* in 1995 and 1996 and projections for 1997 (calendar years for libraries, fiscal years for members) shown in Table 2.

Because the APS will have put all of its journals on line before the end of 1997, with technologies and functionalities that differ from those originally envisaged and implemented for *Physical Review Letters*, it has been decided to treat the journals as being in "beta testing" for libraries for all of 1998. Guideline 4 is therefore being relaxed and the on-line versions will, in fact, be made available to print subscribers – and only to print subscribers – for no additional charge. The prices of print subscriptions will **not** be raised to compensate for this change. Members will be able to subscribe to the on-line versions independently and of print subscriptions, for a redefined and lowered "last copy" price of $25 per journal.

These policies and prices follow the print-based model of selling an annual subscription **to a journal**, for which the subscriber receives', at regular intervals, access to all the articles that are published in an issue. The price is the same regardless of how many times an issue or an article is accessed or read or what the potential is for access. A library with thousands of potential users pays as much for one subscription as one with few users. At least this has been true for the first "simultaneous access" subscription. Additional on-line subscriptions have been much less expensive than the first. This departure from the practice for print subscriptions was meant to be a step towards differential pricing depending on potential use. With the pricing scheme adopted for 1998 and the difficulty of controlling and monitoring simultaneous access, this approach has also been suspended.

TABLE 2. Physical Review Letters prices and subscriptions

Item	1995	1996	1997
Members			
Price of print subscription	140	145	150
Number of print subscribers	3845	3278	2900
Price of on-line subscriptions	N.A.	75	75
Number of on-line subscribers	N.A.	857	736
Price of CD-ROM	N.A.	25	5
Libraries			
Price of print subscription	1580	1820	1970
Number of print subscribers	2441	2330	
Price of free standing on-line subscription	N.A.	1700	1850
Price of on-line subscription with print subscription	free (6 mos)	250	250
Price of on-line subscription after the first	160	250	250
Price of CD-ROM	free with on-line	free with on-line	free with on-line
Number of on-line subscribers	454	212	200

It is easy to see that, for on-line distribution, other forms of packaging and pricing could be made available and might be preferable. They range from charging each time an article is accessed – a scheme known libatiously as "paying by the drink" – to licensing unlimited access for one journal or even for a group of journals or all the journals of a publisher for a fixed fee. The licensee could be a university or a university system, a commercial company or a consortium of companies, a national laboratory or all national laboratories, a state, or even a country. A few publishers are beginning to experiment with such schemes.

APS is contemplating similar approaches. No matter what scheme or combination of schemes is adopted, it must bring in revenues to cover the costs of publishing and be fair to and affordable by the consumers. The challenge of arriving at such a solution confronts all publishers of scientific journals, as well as their authors, readers, and library clients. A solution may be several years in coming. During that time APS will make all its journals available for distribution on line, but may sell few electronic subscriptions and hence collect little revenue. We will thus have to rely on the income from a diminishing number of print subscriptions to keep the journals going. The same is likely to be true for other publishers. APS, whose sole mission it is to advance the knowledge and diffusion of physics, expects to be in the forefront of the transition to electronic publishing. If APS cannot meet the economic and cultural challenge, it is unlikely that anyone else will.

Note

The views expressed are those of the author and not necessarily of the American Physical Society.

RIGHTS MANAGEMENT
IN ELECTRONIC ENVIRONMENTS

D. ARMATI
InterTrust Technologies Corp.
460 Oakmead Parkway
Sunnyvale CA 94086, USA
armati@intertrust.com

1. The challenge

How will rights in general and copyright in particular be managed in the global information society? Will the law prove sufficient? Will comprehensive technical solutions be essential to provide a sound foundation for electronic rights commerce? If so, how will the existing players shape these foundations and what might the resulting commercial infrastructure look like?

Finding answers to these questions remains a high priority for those charged with building shareholder value in the world's great and small rights-oriented companies, for users of their products and services and for those who create and add value to those goods and services throughout the value chain, including those in the academic and scientific sectors.

2. The big picture

On the global stage substantial diplomatic progress has already been made towards providing the legal framework for the global information society. In the intellectual property field, the two new treaties negotiated recently at the *World Intellectual Property Organization (WIPO)* conference and the earlier TRIPs Agreement reached under the auspices of the *General Agreement on Tariffs and Trade (GATT)* (now the *World Trade Organization [WTO]*) have been particularly important. Consensus activities conducted by the *Organization for Economic Cooperation and Development (OECD)* and standardisation efforts in the *ISO/IEC*, the *Internet Engineering Task*

Astrophysics and Space Science **247**: 133–144, 1997.
© 1997 *Kluwer Academic Publishers.*

Force (IETF), the *World Wide Web Consortium (W3C)* and other specialist bodies also help to support the diplomatic agenda.

It is to be hoped national legislatures will accede to the new WIPO
treaties in the near term and quickly modify existing copyright laws to deal
with the new technologies.

At a political level there is growing acceptance of the need for a level
legislative playing field, preferably at the international level. This requirement has been well expressed in the EC's Follow-up to the Green Paper on
Copyright and Related Rights in the Information Society.

In this context it is important to note that considerable care needs to
be taken to ensure treaty and legislative language is not too closely tied
to existing business and technical conditions, especially where transient
technical limitations may be inhibiting the ability of market players to adequately reflect their real-world business models in electronic space. Better
to maintain technical neutrality.

In any event, there is now the real prospect of a relatively interoperable international intellectual property legislative infrastructure emerging
within the next few years, including general support for the use of technical
devices and electronic rights management systems to protect the interests
of rightsholders.

How best to build the commercial infrastructure that makes best use of
this support? How best to ensure the most comfortable intellectual property
trading environment?

3. The existing system

One significant hurdle in reaching consensus on the shape of this commercial
infrastructure is the historical baggage of the existing system, constructed
as it has been on the foundations of an earlier industrial paradigm.

The organizations founded during this period to manage intellectual
property rights have grown in parallel, in an almost symbiotic fashion, with
the various analogue media.

So today these organizations are being forced, some distinctly against
their will, to deal with issues raised by much trumpeted convergence of
information and communications technologies. The walls of the old order
are not crumbling altogether. Rather, through the use of new technologies,
the world's clever organizations are metamorphosing (morphing in contemporary terminology) into new forms, forms founded on understanding and
managing end-to-end processes rather than simply piecemeal tasks.

4. Co-opertition

One of the features of the new convergent order is the clear need for some degree of pre-competitive cooperative action (so-called co-opertition) amongst commercial and non-commercial groups to build on the recognition that the management of rights in electronic space is better viewed as an end-to-end process, rather than merely a collection of unrelated tasks facilitated by a series of often incompatible software components.

What is needed is the necessary rights management infrastructure to service the needs of the widest possible range of vertical markets, providing efficient links between those markets. Managing intellectual property rights across these verticals will be one of the more important applications of the technology.

No matter how much the various players may wish to maintain the status quo, the reality is a revolution is well underway. It is transforming the economic models of the past, stripping away layer after layer of unnecessary cost in a relentless quest for greater efficiencies in the dissemination and collaborative and competitive uses of information. This is affecting both public and private undertakings.

In the process of slashing these costs it seems inevitable that, in the absence of protective action, inefficient organizations will disappear, to be replaced by those that can match the needs of the market. Creators and users alike (and more invidividuals and organizations find themselves in both roles), empowered by readily available knowledge of the potential cost savings implied in the wider use of information and communications technologies, will almost certainly withdraw support from those organizations that fail to use these technologies to best advantage, especially to streamline management of rights in digital environments.

The cooperative development of a common, workable platform for the largely automated, low cost trading of rights in all digital environments seems a most important project. Although the outcomes of such a project may presage an eventual change in the character of the rights management industry it must be vigorously embraced. That said, it is important to note these radical changes will take many years. Given that over two-thirds of the world's population have yet to make their first phone call it seems likely there will be an ongoing need for less automated rights management systems for a long time to come. Old and new technologies will need to co-exist.

5. Convergence and fragmentation

Sadly, cooperation is not inevitable. Convergence is also producing some unexpected market fragmentation. Digital technologists are bringing an extraordinary new range of entertainment and information appliances to market. Users of these devices are already showing the diversity of their taste for highly individualized IPR-based products and services which provides some new niche market opportunities but simultaneously causes headaches for those seeking to make profits and manage rights across this dramatically altered technological landscape.

Maximizing the value of intellectual property and managing IPR-based assets in the chaotic technological environment created by the parallel phenomena of convergence and market fragmentation is made even more difficult by the apparent desire of some politicians to tamper with the legislative framework underpinning these new markets.

As a greater proportion of our socio-economic lives are conducted in electronic space we can anticipate a growing political interest in trying to shape the new digital landscape. The structural impact of this new platform for the trading of economic and social rights in general will become more profound over time. Influencing the design and functionalities of the platform may come to be seen as being akin to framing new constitutions.

From the perspective of owners and users of intellectual property rights the design of the platform is also of vital concern. The various efforts underway to codify real-world commercial relationships as a preamble to the digitization and electronic trading of a greater proportion of intellectual property rights-based assets will help to lay the foundation of a common approach capable of supporting the complexities, subtleties, dynamics and sophistication of existing real-world arrangements. A smoothly interoperable, if not uniform system.

6. Electronic delivery over networks

As convergence and fragmentation become even more powerful forces for change, so the technologies involved in the electronic delivery of intellectual property based products and services will alter.

Today's networks are relatively slow and cumbersome. Internet Service Providers in general do not provide a 100% reliable service (indeed Web-TV needs to have four local ISPs for each user location in order to guarantee a connection every time). Content aggregators are still learning what their customers will use and how to make profits. Hybrid on-line/CD-ROM technologies are almost certainly just a staging post on the path to ubiquitous light-speed networks.

In time, the current offerings will give way to a magical plethora of new devices and services. We are already starting to see some of these. TV-PC (Web-TV etc.) delivery is only the beginning. Remote, handheld, mobile, anywhere, anytime devices and matching services will conspire to redefine the concept of 'the desktop' and potentially the nature of work itself.

Every conceivable part of the electromagnetic spectrum will be utilized to provide such services. Issues of reliability and availability should retreat into the background as bandwidth and network intelligence expand.

Will these networks always remain neutral to IPR-based content carried? Not at all. As more of our economic lives are spent in networked electronic space so the value of 'information exhaust' – information gathered both on individual transactions and in the aggregate – will become ever more valuable. Indeed some business models may be based exclusively on this revenue source. Questions as to who owns the rights to the 'information exhaust' will become an important part of commercial negotiations.

In general, network operators themselves, having fought vociferously against the proposition, are now unlikely to be held legally responsible for IPR management. This does not mean, however, that the network operators will be disinterested in adding value to their services by providing mechanisms that enable rightsholders to look after their interests more or less automatically. It is perfectly feasible to imagine telecommunications operators banding together to provide helpful global solutions in this field either for the public network, or more likely in the short to medium term, private networks.

7. Superdistribution

Until now true 'superdistribution' of intellectual property assets has not been feasible. In order to maintain control over these assets it has been necessary for clients to be in communication with a server. With entirely new technologies (such as InterTrust Commerce Architecture™) now coming to market this is set to change. Using these technologies the threat posed by a customer passing along valuable intellectual property to friends and colleagues, so familiar from the existing analogue world, can been turned into a business opportunity.

Once rightsowners realize that they can persistently secure their content, enforce business rules relating to its use, including payment, it is likely that client-server distribution models will begin to give way to peer-to-peer systems. This would have significant beneficial cost implications for distributors of intellectual property assets as a greater proportion of the overall distribution cost would be borne by users.

Whether or not it is existing publishers and other producers of intellectual property based goods and services who take advantage of these new opportunities or a new set of service providers emerges, there seems little economic reason why the superdistribution model should not, in time, succeed.

8. Further implications

One of the most difficult technological challenges has been to develop ways to enable the great diversity of real-world commerce to take place in electronic space.

As the first solution to address these issues head-on, the InterTrust Commerce Architecture, principally based on DigiBox™ secure containers, InterTrust Commerce Nodes and a secure operating system layer extension, enables the creation of a globally distributed Intertrustworthy™ environment. This environment provides the basis for the broad range of real-world rights and obligations, including intellectual property rights, and real-world rules related to those rights and obligations, to be securely expressed and enforced.

Use of this distributed trust environment means that dynamic, agreement-based relationships can be treated as 'trusted' by the parties and managed across time and electronic space, in all kinds of electronic devices, from supercomputers to hand-held personal digital assistants, smart cards and even simple household appliances.

One of the strengths of the technology is that the rights to use of the contents of an object can be managed separately from the contents themselves. The rights to the underlying works (the original creation) can be managed separately from the rights relating to particular expressions of those works, just as they are in the real world.

Transposing real-world IPR-based business practices to the electronic dimension Many relationships involving IPR-based assets are framed and managed by contractual agreements.

Significant work still lies ahead in codifying and digitizing contractual terms to enable all the richness of real-world agreements to be reflected in their electronic equivalents.

Changes in the parties with interests in the rights to a particular digital object and the terms and conditions of agreements amongst these parties must also become more elegantly manageable. For such changes to be made simply and efficiently it is essential to have a distribution architecture that provides for separate delivery of the terms and conditions of the current rights management agreement and the contents to which the agreement relates.

Another important feature of rights and contracts is time. In order to maintain a trustworthy distributed environment for electronic information commerce it is vital that a persistently reliable source of time is available. For the management of various forms of time-based intellectual property rights this is particularly so.

Techniques also need to be used that ensure integrity and authenticity of digital objects. Not only do these increase the overall reliability of the system, they also enable management of some forms of author's rights. These components are now also readily available.

9. Scale and scope

Further important IPR issues generated by network delivery are scale and scope. Managing IPR globally in electronic space, with all its nooks and crannies, is a truly mammoth task, in large measure due to the differences (some minor, some not) in national intellectual property laws.

Designing IPR management systems that scale to cope with billions of users and trillions of digital objects is a significant problem. Creating clearinghouses capable of efficiently processing payments resulting from transactions related to the rights associated with these objects also requires the involvement of very sophisticated companies. To effect trusted transactions, objects must be reliably identifiable, as must the current rightsholders and all other interested parties in the value chain.

Cross-border and at-the-border issues also present potentially costly complications. Territorial matters – international, national, regional and local, the subject of tensions the real world, are already showing their capacity to rock the boat in electronic space also. Cross-border and at-the border issues have relevance, too, for the enterprise in its relationships with other enterprises.

Legal questions relating to jurisdiction, taxation and currency are gradually being addressed, but the final picture is still very far from clear.

10. Maintaining diversity

A significant challenge for IP rightsholders may come from those who would wish to see digital technologies used to underpin an entirely new approach to the sharing of creative output. Early public users of Internet technologies are renowned for their stinginess.

Many industry leaders and commentators take the view that internet distribution of a large proportion of IPR-based information products will be supported by advertisers rather than consumer micropayments – a free to

the user broadcasting model rather than a pay by the slice micropublishing
model.

While this view may remain true in the short term it will almost certainly be eclipsed by future developments.

As the market matures it seems highly probable the fragmentation factor will lead to the need to reflect all the complexities and subtleties of
real-world arrangements in the electronic marketplace as well as new models built on the foundation of a networked distributed trust architecture.

Just as technologies need to be advanced that support these, so do we
need to ensure that political support remains strong for a sensible mix of individual and collective control of intellectual property rights. To unbalance
the scales in favour of one model or another would be extremely unwise.

11. Do users care?

In general terms, users know little and care even less about intellectual
property rights. It was disturbing ,yet not surprising, to read the Report
of the Electronic Copyright Survey 1996, prepared by Benchmark Research
for The Investext Group. This report showed amongst the European information professionals surveyed 'there is a low level of knowledge regarding
copyright issues in general'. 82% of respondents also find it 'difficult' or
'very difficult' to keep up to date with these issues.

This has not inhibited the development of great industries based on
copyright.

Users do care about ease of access to copyright-based products. Those
systems which provide poor transactional interfaces will inevitably drive
away business or lead to piracy. If every time a consumer wishes to use a
tiny component of a multimedia product they have to complete a cumbersome transaction it seems highly probable they will not support the general
system underpinning this approach – driving technologists to draw the conclusion that the most sustainable model is free to the user, supported by
advertising. Like commercial television and radio, this model obviates the
need for users to negotiate complex transactional systems – just turn on
and tune in.

Similarly simple interfaces, with sophisticated rights management being
handled automatically in the deep background, need to be developed to
streamline desktop information delivery and collaborative systems.

Our physical world is full of simple ways to be parted from our money.
On the back of increasingly ubiquitous electronic systems and the pressing
competitive need to reduce transaction costs the barriers grow ever smaller.
Contemporary electronic transaction systems are the product of several
iterations involving commercial, marketing and engineering input in large

doses. By bridging manual and electronic systems they support a dazzling array of business practices.

Surely models operating entirely in electronic space should be able to provide the most delightful, least obtrusive, simplest to use, least costly transaction systems. Before long, with close reference to customers, they will. For this to occur several components of real world commerce need to be available to designers of end-to-end electronic information commerce systems. These include low cost ways to guarantee identity, integrity, authenticity and privacy, automated payment to all members of the value chain and repudiation.

12. Balancing the social cost: benefit of rights management

Clearly enormous social and economic benefits have accrued on the basis of intellectual property rights. Substantial (and growing) proportions of global GDP are generated by IPR-based industries.

Faced with reconciling political pressure from various interest groups, from government agencies, from the IPR-based industries themselves, and from the myriad users of their products and services, it may be at times difficult for politicians to avoid drafting unbalanced legislation. Once held up to public scrutiny, however, common sense is the usual winner. Already there have been active (and largely successful) lobbying campaigns mounted to defeat measures considered to tip the scales in favour of one or other of these groups.

Some values will not change. Leisurely public access to the knowledge base of humankind, as afforded by the public libraries of the world for the past century, is one of these. So, too, is the social desire to protect the interests of particular groups of users by enabling differential pricing (including free access and use) based on group membership. Privacy is also of paramount concern in many countries and to many individual and corporate users.

It is important for legislators to know that technologies now exist that can underpin these values in electronic space – protecting the legitimate interests of both rightsholders and users alike, in some respects even more elegantly and efficiently than they are in the existing hybrid physical/electronic world.

Another political IPR issue involves the liability of internet and on-line service providers for IPR management within their systems. Although it seems in most jurisdictions they may not be held legally responsible, they appear likely to be increasingly held accountable politically for selectively screening the types of content to which they allow users to have access. This may have potentially deleterious effects on IPR-based industries. Technologies that enable automatic compliance with various political directives may prove useful.

In intranets the reverse is likely to be true – controllers of intranets will likely be held liable for management of intellectual property originating from within and introduced from outside. They may not, however, depending on the context, be held directly accountable, except by their internal constituents, for the types of content carried.

Interactivity and in-line relationships: IPR in collaborative information partnerships

In the past, for the most part, creators were creators, producers were producers, distributors were distributors, users were users, rights managers were rights managers and bankers were bankers.

New interactive technologies are blurring these distinctions. In intellectual property terms this is important. In an interactive information exchange who creates value and who consumes it?

The levels of interactivity enabled by complex software products and internetwork technologies are shifting our sensibilities. We are moving from an on-line world, in which a dependent workstation was linked to a omnipotent host, to an interactive client-server model. The next stage will be the evolution of a peer-to-peer, in-line world in which relationships are far more collaborative, information partnerships far more fluid, with all involved (large and small) having the opportunity to become participants in the information value chain.

How do we effectively manage these in-line relationships? How do we attribute authorship and allocate revenue streams? We need to have ways in which to adequately reflect this new reality. Again, fortunately, solutions are now becoming available.

13. Information exhaust

Another important aspect of the widespread use of internetworked interactive information systems is the opportunity they afford to gather very precise information on user behaviours. So called 'information exhaust'. Interest in this information is intense. For many IPR-based businesses it may become a primary source of revenue.

How this information is gathered and used will almost certainly become the subject of hot debate. Internal managerial use of exhaust generated as a result of interaction with information products within the confines of an organisation may be less controversial than the gathering and sale of usage data generated as a result of user interaction with sites external to the organization. It seems desirable that any rights management system also support the full range of privacy options – from complete privacy to complete openness – in a relatively standardized fashion.

T'here are already indications that privacy will become a divisible, tradable notion. We will be paid, often in the form of discounts or free access to particular electronic services, to surrender it in whole or part.

14. Is the market efficient?

As the potential of peer-to-peer systems becomes clearer, market intermediaries of all kinds will have to demonstrate real value-adding input in order to have an economic role. Channel control will generally become looser. Market fragmentation will add to this lessening of power concentration.

Channel control may rest more with the control of technologies (such as set-top box decoders, and personal smart card readers) rather than on control of market channels in the traditional sense.

For intellectual property rightsholders and users the best approach is to be actively involved at the leading edge in the design of comprehensive, end-to-end solutions that meet as many of their interests as possible, rather than waiting for others to deliver proprietary turn-key products that may limit their bargaining power and ability to change direction with changing market conditions.

Discontinuity as a new industry finds its feet: another 15 years of red ink?

At the recent MILIA'97 in Cannes, the multimedia content industry asked itself whether it faced another fifteen years of red ink before finally becoming a mainstream, profitable industry. The same might be asked in all those industries that have been backing internet technologies to provide the basis for new and exciting global businesses.

Is internet technology merely providing a new channel, one of many on the new hybrid PC-TVs? Or does it have the capacity to radically alter existing industrial structures? If so, what might the new structures look like?

The answer (in part) may be that there will be winners and losers and the winners will determine the shape of the new structures. Those market leading corporations and non-commercial institutions who enjoy dominance of particular vertical markets could successfully invest in technologies that perpetuate their dominance in electronic space. Some will, some will not. New players will undoubtedly emerge. Entirely unpredictable events may completely alter apparently logical outcomes. One thing is certain, however, the phenomenon will not go away and it will provide the basis for an expanded set of business opportunities.

15. Embedding societies in silicon and software

As we proceed with the process of embedding our societies and their values in silicon and software, our goal then, should be to enable the full spectrum of real world relationships to be expressed and managed reliably in electronic space.

In order to make this a reality a fundamental requirement is to build an end-to-end environment in which distributed trust is a given – where we can rely absolutely on the underlying architecture to provide a trusted foundation on which to build the processes that facilitate our normal day-to-day existence.

16. Risks – Opportunities – Carpe diem!

There are undoubtedly risks and opportunities aplenty. This is not, however, a time for timidity. At every level of the value chain we must seize the day and participate in this profoundly important task of designing and building a trusted rights management layer as a key foundational element of the technological platform for a globally acceptable information society.

Such a platform will enable participants throughout the information value chain to vigorously compete with one another on the really important aspects of their value propositions – matters such as the quality, timeliness, depth and individualized nature of the information content they offer.

The era in which it has been adequate to deal with rights management issues as a series of piecemeal tasks is drawing to a close. The age of a comprehensive end-to-end, process-oriented solution is about to dawn. For those charged with ensuring copyright compliance in ever more globally distributed systems, whether as rightsholders or as users of information, this shift will be a very welcome relief indeed.

PROTECTING PUBLISHED SCIENCE

E. BARROW
UK Copyright Licensing Agency
32 Plaid Road
London SW2 5UR, UK
edward@cla.co.uk

"Science is not science until it is published". Publication of the reviewed and refereed article is the end of the science process. In blue-sky disciplines, the published paper is almost the only quantifiable result. A cynic might wonder at the wisdom of spending so much money – particularly in high-energy physics and astronomy – to produce a few sheets of paper. But because so much investment has gone into their production, those few sheets of paper – or few kilobytes of text – deserve protection.

The paper serves a number of important functions. Firstly, it records the achievements of the scientist. Secondly, it informs other scientists – providing them with the information they need further to pursue science. Thirdly, it records scientific progress (and forms the archive of science). Fourthly, it informs the world – including scientists from other disciplines. To fulfil these functions, the paper must be processed in a way that requires specific skills, not necessarily scientific skills – which are normally provided from outside science. Publishers and librarians are involved. For a range of historical and pragmatic reasons, publishers come from outside the academic community whereas librarians are part of it. Librarians thus receive direct public or foundation finance, but publishers rely on commerce to make a living.

What the publisher does has changed over time. Until the 1980s, the most obvious service provided by the publisher was to finance printing, and in particular the substantial outlay required for typesetting – but it was far from the only service. Scholarly papers have to be reviewed prior to publication. Managing the peer-review process and ensuring its rigour and impartiality has now become the publisher's most important rôle. However, whatever the relative importance of the various services provided by the publisher, they all cost money and the publisher's investment must be recouped through a range of commercial charges. These are traditionally

Astrophysics and Space Science **247**: 145–153, 1997.

journal subscriptions and page charges, and both are paid out by the public sector to the publisher. Copyright is an important legal mechanism used by publishers to protect their revenues. Unfortunately there is no equivalent mechanism available to publicly-funded academic institutions and libraries to protect their revenues from the caprices of their political paymasters.

Until the Second World War, scholarly publishing had been a relatively small-scale activity at which no one had made much money. The contribution of science and technology to the outcome of the war resulted in a huge expansion of science and a consequence expansion of scientific publishing. New entrants to the publishing market, in particular (but not only) the late Robert Maxwell, realised the scope for large profits to be made by increasing subscription charges given the then prevalent inertia of librarians. Library budgets were also stretched by the growth in the number of titles reflecting the growth in science itself – and not merely for reasons of economy, the photocopier came to be used as a tool to give academic faculty wider access to the burgeoning range of scientific literature. When governments started to curtail the growth in their expenditure on science, libraries responded by introducing cancellation programmes and publishers responded by enforcing copyright, through the development of photocopy licensing. In some sectors of the academic community, copyright came to be seen as a tool for extortion.

Against this background, the 1990s are bringing a change which is potentially as significant as the invention of the printing press. Publishers once again feel threatened, and once again are inclined to turn to copyright to protect their interests. And there are voices within the academic community calling for an end to copyright. The dramatic change of the 1990s (though one with much older roots) is the development of networked electronic information distribution: in particular, the Internet.

Copyright is a legal doctrine which evolved in response to the technology of the printing press. Printers made substantial investments in producing new books, which was a risky business. In Europe in the sixteenth and seventeenth centuries, the caprices of kings and governments were perhaps even more unpredictable than those of the book-buying public, and in the midst of the religious and political turmoil of those times booksellers had to tread a careful path between the tastes of the public and the authorities. When a book actually found the right balance, other printers would produce a rival edition – incurring the costs of typesetting only when confident in the commercial success of the book. The printers formed themselves into guilds and the guilds made a Faustian pact with the authorities: given the right to regulate competition between printers, they would ensure that nothing seditious was printed. The printers' rights controlled by the guilds were called copyrights. It was a system to limit competition and implement censorship, and by the end of the seventeenth century in Britain it had become anachronistic. Parliament, not the monarchy, was the dominant political force; and as trade had expanded the guilds had lost much of their control.

In a stunning piece of legal innovation, the old copyright system was transformed into one that is recognisable today – by making copyright a fundamental right of the author, not the printer or bookseller, and removing entirely the link with censorship. At last, the rights of the author were recognised. The associated innovation was to make the copyright a property right which could be transferred – for example, to a printer – in consideration of something – for example, a sum of money or a continuous stream of royalties. Modern copyright evolved not just in response to the printing press, but also to the declining power of patronage.

Copyright today is an international system, but in some countries it is viewed with intense suspicion. Much of this is due to the almost imperialist vigour with which the doctrine is espoused by those countries which are significant exporters of copyright material – music, films and computer software – of which the United States is the prime example. Although most Asian countries now belong to the international conventions on copyright, it is explained that it does not sit happily with their traditions.

The writer or scholar and the artist is supposed to work purely for the love of the work, and to earn not money but respect. Buddhist and Confucian traditions require that the wealthy support scholars and artists through systems of patronage; Confucius himself depended throughout his life on the patronage of the powerful. But these ideas are not as Eastern as they seem: until the development of copyright, art and scholarship in the West was similarly dependent on patronage. Indeed, today, scholarship is dependent on patronage: and no scholastic discipline more so than high-energy physics with its demands for ever bigger and more powerful particle accelerators, and ever bigger and more accurate telescopes. And just as many Asian and developing countries only grudgingly accept the principles of copyright, so too it is often a problem for academics and academic institutions.

It is perhaps reasonable to ask whether the reason for the lack of broad acceptance of copyright both by developing countries and by academics stems from the fact that it is perceived as a tool for extortion. The suspicion may even be based in reality. Nevertheless, the fact that copyright is sometimes abused does not alter the fact that it is a fundamentally just principle, based on democracy not patronage, and on individual accountability not government censorship. The basic natural justice underlying copyright stems exclusively from the fact that the right belongs first to the author. With this premise always in mind, copyright is as relevant to the technology of electronic publishing as it is to the printing press; and it is as relevant to the academic author depending upon latter-day patronage as it is to the novelist in his garret. Because the author owns the rights initially, the author can dispose of them at his or her discretion.

The rights assigned to the publisher "in consideration of publication" are very valuable. That is not to say that the scholarly contributor to a learned journal always gets a raw deal (although those authors who are paid directly for what they write are horrified by the practice). To a scholar, publication is a valuable reward and the assignment of rights in a paper is a small price to pay. Once assigned, they are for the publisher to use and for the publisher to protect. However, let us suppose that the author retains the rights. Is protection still desirable?

Protection against piracy is perhaps less of an issue to the scholar. The scholar's objectives in publication are most honourably the dissemination of his or her work, and most realistically the career benefits of the imprimatur of publication in a leading peer-reviewed journal. The more it is copied, the more the author benefits – as it is more widely read, so it is more likely to be cited, so the scholar's reputation grows, so the likelihood of funding for the next research project increases. The publisher is equally concerned to see that his or her publication is widely read, but on condition that it is paid for. One form of protection ensures that those who have not paid (or do not belong to an institution which has paid) may not read the material. This form of protection benefits the author insofar as it maintains the economic value of the work that has been published - but in that it may impede free access to the works of peers useful for the furtherance of research, it is a disadvantage.

Economic protection of the material is much less important if the publisher's investment in the publication, editorial and review process is recouped by page charges alone rather than subscription. There is strong support amongst some sectors of the science community for this approach, in which the cost of publication of the completed research is borne by the research project but once published the information is freely available. In other areas of the publishing industry, the principle of payment by the author is known as "vanity publishing" and is regarded with contempt. In scholarly print-on-paper publishing, page charges have traditionally been espoused primarily by US learned society publishers (and even so do not replace subscriptions entirely); whereas learned societies in Europe, and commercial STM publishers worldwide, have favoured subscription-only payment. The threat presented by a page-charge system of payment is that it could undermine the impartiality of the publisher's selection process.

Certain pharmaceutical companies, for example, do not permit their research scientists to publish in journals which levy page charges for fear that they could be perceived to be using their financial muscle to influence the publication decision. However, it is equally possible that the publisher dependent on subscription income would be tempted to publish popular (or controversial) rather than meritorious papers. This is an argument put forward by some for removing scholarly publication entirely from the commercial arena and placing it solely in the hands of scholars themselves - either learned societies or a new publishing enterprise owned entirely by scholastic institutions. Disaffection with commercial publishers is nothing new; in the United Kingdom at the end of the sixteenth century the universities of Oxford and Cambridge established their own printing presses when the London stationers failed to meet their needs. Despite still being an integral part of their respective universities, in matters of copyright OUP and CUP nowadays ally themselves closely with the commercial publishers. Scholarly publication ventures established under similar pressures in the electronic environment are likely in the fullness of time similarly to change sides in the copyright debate as the financial issues come to dominate their management's thinking.

It is in the interests of scholars that the publishing process should remain economically viable. If a subscription model for electronic publishing is chosen, some form of economic protection becomes highly desirable. However, whichever model is chosen, another form of protection is also required. Electronic media are vulnerable to alteration as well as to copying. Alteration can be both deliberate and accidental (as a result of a transmission error); in either case, the paper as read by the recipient is not the same as the one written by the author. The integrity of the author's paper is then a matter of concern.

Current protection technologies address both economic protection and the protection of integrity. There are three main approaches to the problem of protection, which may be summarised as: tattooing, fingerprinting and encryption.

Tattooing (or watermarking) involves the insertion of an identifying signature deep within the structure of the electronic file. The identifying signature is contained in the data itself, and attempts to remove the identifier also degrade the data unacceptably. Many implementations of tattooing technology survive digital/analogue/digital conversion. The tattoo helps economic protection by enabling usage to be tracked and appropriate charges to be levied. However, the technology has some disadvantages. Firstly, it relies on digital files being essentially bit-mapped representations of analogue data, in which there is inevitably some redundancy. The redundancy is used to record the tattoo. Sound, images and video are almost invariably carried in a bit-mapped format, but there are much more efficient ways of representing text. In an ASCII text file, there may not be sufficient redundancy to carry the tattoo. Tattooing is unlikely to work with pure HTML – (the language of the World Wide Web), because HTML files are nothing more than ASCII text files. This does not meant that tattoo protection cannot be used on the World Wide Web, but that it can not be used to protect the underlying structural features of the Web.

Secondly, tattooing does not prevent unauthorised use, but merely provides a facility for use to be recorded. The economic protection is thus not particularly effective.

Fingerprinting is an extension of tattooing, in which the tattoo is written to each time the file is accessed with the identity of the user or process which accessed the file. Thus a trail is left which makes the detection of unauthorised use simpler. Fingerprinting technology is in its infancy and many technologies which are described as being fingerprinting are, upon further examination, merely watermarking or tattooing technologies.

The third method of protection is encryption. It is a very powerful technology which potentially offers very strong economic and integrity protection of any digital data. It faces, however, many regulatory problems. The primary use of encryption is for the preservation of privacy. Very powerful encryption algorithms for privacy protection are now in the public domain; these include the US Department of Defense's Data Encryption Standard, or DES, and the Rivest-Shamir-Adelman public-key/private-key algorithm. The latter, in particular, has caused much controversy since it is classed by the US government as a munition and is subject to export controls. Encryption technology is militarily sensitive: we are only now becoming aware of the major rôle that the successful Allied deciphering of the German Enigma code played in determining the outcome not just of the Battle of the Atlantic but ultimately the second world war. The Enigma machines were highly secret, and the Allied task in deciphering the code was greatly simplified when one of the machines was captured. By contrast, modern encryption algorithms are not secret. The only secret part is the private key, but even with full knowledge of the algorithm and the public key, it can be mathematically proven that RSA encryption cannot realistically be deciphered. Security weaknesses in RSA arise from flaws in its implementation rather than its intrinsic structure.

Unfortunate as it may be to the authorities, encryption technology is widely available. Philip Zimmerman's Pretty Good Privacy, a program which implements the RSA algorithm, is freely downloadable from a large number of web sites. If there were to be war, strong encryption would now be available to all sides. In peacetime, the authorities are still concerned to be able to intercept the communications of citizens, alleging that it is vital if they are to prevent terrorism, for example, or drug smuggling. Nevertheless, the demand for secure private communication – for example, to enable commerce over the Internet – is growing. The current policy of the US government is to restrict the export of strong encryption, while giving the software manufacturers two years to develop a system of encryption with a back door through which the law enforcement agencies could monitor communications. Even if successful, this will lead to the position where law-abiding citizens have no privacy whilst the terrorists and drug smugglers will continue to communicate using existing secure technology such as PGP.

Other countries adopt even stronger rules on encryption. In France, for example, its use must be licensed by the government and licences are very hard to obtain. As contributors to this publication, we were warned not to send our contributions in an encrypted format to the editor in France, because of the illegality of unlicensed encryption.

In its basic form, encryption prevents the first unauthorised use of a protected work, but once decrypted the material can be retransmitted. Encryption-based copyright protection technologies overcome this problem by permitting decryption only into a controlled environment, established on the reader's workstation, from which material cannot be saved or transmitted. This method provides for secure economic protection. Integrity protection is integral to encryption as a result of the mathematics of encryption; the encryption software can be used to generate a unique code based on the original digital file. Any change, however minor, in the original file, will result in a different code being generated so that the integrity of the original can be guaranteed.

Public-key/private-key encryption technology also allows the most powerful proof of authenticity. This method of encryption relies on asymmetric functions. Keys exist in pairs, and a file encrypted with one key can only be decrypted with its matching pair. The normal procedure is for each user to have a pair of keys, one of which is kept secret while the other is publicised. A recipient of an encrypted file which can be decrypted with a particular person's public key can be confident that it was encrypted with the matching private key – and thus of the origin of the file.

The regulatory problems associated with encryption are considerable, but if they can be overcome then the benefits of the technology to scientific publication are at least as great as those to electronic commerce. Overcoming the regulatory problems is primarily a matter of convincing the security agencies of governments that this particular horse has already bolted, and no matter how tight they try to close the stable door, the horse will remain on the outside. The extent to which governments are unwilling to accept this fact is perhaps a worrying indication of the extent to which their law enforcement strategies depend upon being able to intercept communications.

Proof of integrity using encryption-related technologies is desirable. It has, perhaps, further potential in scientific publishing. So far, the discussion in this paper has concentrated on the publication of the results of science – the refereed paper. The paper, as its name implies, is a relic from the days of print-on-paper publishing. Electronic publishing offers far greater potential. Words are not always the best way to describe complex scientific concepts; indeed, much of science (and certainly astronomy and high-energy physics) relies on the language of mathematics to express ideas which cannot be expressed in words. Other scientific concepts depend upon other representations, such as diagrams: but diagrams themselves show their weakness when illustrating, for example, the physics or mathematics of higher dimensions. In chemistry, the use of 3-D molecular modelling software is becoming widespread, and this particular representation of chemistry is at least as appropriate as words on paper. Electronic publishing technology has the potential to be used creatively by scientists to develop means of representing and explaining science other than the conventional paper of words in a row. In particular, there is potential for the emphasis of publication to move from the results of science to the process of science.

Publication and peer review is the mechanism science uses to maintain standards. Regrettably, in a very few cases, scientists have been known to falsify their data for fraudulent purposes. In physics and astronomy, this is perhaps less likely than in other disciplines, because the capital-intensive nature of equipment necessitates more collaborative research. Nevertheless, unprotected the digital medium lends itself to fraudulent falsification. Scientific equipment increasingly produces its raw output in digital form. Where analogue output is produced, – as for example photomicrographs – these are often digitised onto PhotoCD for archival purposes. The illustrations to the published paper may be taken from these photomicrographs and digitally enhanced, often for valid reasons. Enhancing the contrast can make subtle distinctions more apparent to the non-expert reader, thus helping with the process of publicising science. But it is just as easy to falsify the data entirely.

This paper is entitled "protecting published science". It may be, however, that the developing digital protection technologies are as well suited to the protection of unpublished science – including, in particular, raw data

– against scientific fraud as to the protection of published science against misappropriation. Electronic publishing and archiving allows the raw data to be associated with (or hyperlinked to) the published paper, thus allowing both referees and interested readers to view the raw data. Publishing raw data has a further benefit: it can simplify meta-analysis, in which the data from various experiments can be used to provide a larger sample.

Electronic publication will only truly revolutionise scientific publishing if it changes what is published, as well as the manner of its publication. The full potential, however, can only be realised if protection technologies are used to protect both the economic value and the integrity of the published material; and this is only possible if the political and security community relaxes its attitude towards the exploitation of cryptography. Scientists are always asking politicians for favours, usually in the form of money – but liberalising attitudes to cryptography does not cost money. The boost to electronic commerce is likely to increase taxation revenues, thus allowing governments to cut less funding from physics and astronomy programmes.

BITS AND BYTES AND STILL A LOT OF PAPER: ASTRONOMY LIBRARIES AND LIBRARIANS IN THE AGE OF ELECTRONIC PUBLISHING

U. GROTHKOPF
European Southern Observatory
Karl-Schwarzschild-Straße 2
D-85748 Garching, Germany
esolib@eso.org

1. Introduction

The future has already begun. The information superhighway, hypermedia, digital libraries and electronic publishing are not vague concepts anymore that might be awaiting us beyond the year 2000. Instead, they are already here. Each day our electronic mailbox is flooded with announcements about new sites on the World Wide Web that vie for our attention. More and more information resources are easily accessible and need to be checked out. Modern communication technology has brought us so close together that all the information available anywhere on this planet seems to be at our fingertips. This is the Information Age.

For us librarians, these are not only extremely fascinating, but also very challenging, times. Librarians have been information providers for centuries, and nowadays we have the opportunity to use tools that allow us to provide an even quicker, more complete and sometimes more sophisticated service to our users. Databases and reference sources can be queried via the Internet, library catalogs are available remotely, newsgroups and mailing lists provide a wonderful opportunity to discuss mutual concerns quickly, and electronic mail allows us to contact colleagues who might be able to help us solve a problem. And all this can be done within a minimum of time.

Technology will take us one step further very soon. It is no longer only references and information *about* documents that is provided electronically, but also full texts of publications that have become available in digitized form. If we look for information about a particular topic today, we can start

Astrophysics and Space Science **247**: 155–174, 1997.
© 1997 *Kluwer Academic Publishers.*

for instance by querying a library catalog on the World Wide Web (WWW). The results we receive might consist of some formal bibliographic data like authors, title, and publication date. Some content-related information (keywords, thesaurus descriptors, or the table of contents) may be included as well. Plus, we may have the opportunity to access the abstract or even the entire text, as well as graphics, tables, video sequences, or an embedded software program that can be run locally. The references at the end of the publication are active links. One mouseclick, and we move on to the referenced paper without even noticing where exactly we are going on the network. Another mouseclick, and a search engine finds other publications dealing with similar topics that also might be of interest to us. Particularly useful documents can be printed or downloaded onto our own computer for later use. The search strategy we chose to find these publications can be stored and reused whenever needed.

Access to information is becoming more and more seamless. The year 1997 is a milestone in astronomical publishing, as some of the most important publishers recently launched electronic versions of their journals or intend to do so in the course of the year. However, various barriers currently impede immediate access to every piece of information. In the following, we will look at the many advantages of electronic publications, but will also focus on some of the questions and unresolved problems.

Change in general, and technological change in particular, has its price. Librarians are worried about some of the current developments and therefore speak up if short-sighted decisions are being taken. Neil Postman, a North American communications theorist, stated in his book "Technopoly" that

> [...] it is a mistake to suppose that any technological innovation has a one-sided effect. Every technology is both a burden and a blessing; not either-or, but this-and-that. (Postman 1992, pp. 4-5)

It is necessary to be aware of both the positive and the negative implications in order to make the most efficient use of technology.

2. From Traditional to Digital Libraries

The tools used by librarians in their daily work have changed vastly during recent years. Today, hardly any library is equipped exactly as it was only a few years ago. In addition to traditional means like card catalogs and microfiche readers, most libraries now also offer an online public access catalog (OPAC), public PCs equipped with CD-ROM drives, scanners, or public terminals connected to the Internet. An increasing number of libraries are building homepages on the World Wide Web from where users have access to a variety of services without physically entering a library.

Many libraries are in transit from the traditional towards the digital library. We witness a shift from libraries offering information about (electronic and print) information towards providing access to full texts of documents. Not only recent publications, but also many historical library holdings are being digitized (see e.g. Corbin & Coletti 1995). These electronic collections allow users from everywhere at any time to consult the material without doing any harm to fragile documents.

Despite numerous digitization projects, electronic media by no means are dominant compared to print material. There is still a lot of paper in our libraries, and we expect this to be the case for a long time to come. The paper-based library will co-exist with the digital library for the foreseeable future, because electronic publications are not developing at the expense of print media, but in addition to them.

The notion of library has long expanded beyond the physical building of the library. Our services always included access to sources that are physically located outside the library. Over the course of the years, librarians have collaborated in many ways. Central cataloging, union lists of journals, cooperative collection development and interlibrary loan are only a few examples of resource sharing. Forced by decreasing budgets, many libraries have redefined their acquisitions policy from purchasing documents "just in case" to "just in time", since no library can afford to purchase every item that might be needed by one user one day. Through collaboration and reciprocal services among libraries, we can provide a much larger range of resources to our users and fulfill their information needs quicker, cheaper, and more completely than one library alone would be able to do.

While projects that aim at helping each other might be seen as a nicety during prosperous years and become a necessity in times of economical restraints, they play an ever more essential role in the electronic environment. James Michael (1994) suggested a blueprint for the library without walls that consisted of five elements:

1. interconnectivity – connecting to a network
2. interoperability – the ability of one computer to talk to another
3. integration – of internal and external resources into one single user interface
4. intermediation – reference services, navigational help and instruction provided by librarians
5. interdependency – because one single library cannot own all the resources that might ever be needed by users

This last item, interdependency, is the final step for the "Global Digital Library" to become reality. In the electronic environment, even more than in the traditional paper-based world, no library can (or may) store all the documents to which it provides access. Digital libraries are only possible if reliable partners cooperate on a long-term basis. Authors, libraries, publishers, archives – the concept of one player in the electronic publishing sector as a self-sufficient entity has been overcome for good. The digital library indeed brings us closer together than ever.

3. Electronic Publications: Impact on Library Functions

3.1. ELECTRONIC VERSUS PAPER-BASED PUBLICATIONS

Without doubt, electronic publications in general, and electronic journals in particular, provide some advantages compared to paper-based documents. Electronic documents typically are delivered with a powerful mechanism to search easily through them. Navigation tools allow readers to jump to particular sections, e.g. to references or graphics, and from there to the relevant section in the body of the text. "Forward referencing" provides links to articles which were published later, but cite the original article. Corrections can be included without difficulty. References can be linked to abstracting services, from where abstracts or full texts of cited papers can be obtained, and similar publications can be retrieved based on the original article. Electronic versions of journals usually are available in advance of printed versions, and in addition users can browse the tables of contents of forthcoming issues. Journal issues need not be shipped, thus avoiding delays due to mailing systems. Electronic documents can be accessed from anywhere at any time and by as many simultaneous users as needed.

New tools will go beyond these features soon. Graphics will be available together with the underlying data, allowing readers to modify them according to their needs; video or sound sequences as well as computer programs can be embedded; journal issues can be linked to data archives, image libraries, laboratory measurements, or software collections (Boyce 1996), or users may be able to launch demonstrations or other virtual experiments.

On the other hand, electronic publications, at least if used with today's technology, are bad for browsing and nearly unacceptable for reading online. Interesting documents are likely to be printed, probably even several times, at the users' sites, thus shifting printing costs from the publishers to the scientists' institutions. Further costs arise because special tools, both hardware and software, may be required for downloading, reading, and using electronic publications. Unless a text-only version of electronic publications is provided, terminals need to have graphical capabilities. Specific versions of Web browsers need to be installed in order to correctly display the documents. Appropriate software might be necessary to store or print electronic publications. Bandwidth is unlikely to keep up with demand at least in the near future (Butler 1996); if network capacities are not sufficiently high, access to documents can be painfully slow or impossible. Even with powerful connections, loading of individual documents on the Web can be extremely time-intensive, for instance if the providers include extensive graphics. Many network users try to work around this problem by simply switching off the display of graphics in their Web browsers.

Costs for electronic publications are not limited to purchasing equipment. The myth of all information on the Internet being free-of-charge has not yet been destroyed. In the first days of the net, many documents were maintained by individuals who spent their private time and personal efforts into setting up network resources. This voluntary attitude has led to the widespread, but wrong assumption that electronically delivered information doesn't cost anything. The truth is that the Internet is just the communication channel; valuable content must be paid for as much as before. Commercial publishers are about to leave their experimental phase during which access to their publications often was allowed without any charge, for testing purposes, and now are establishing subscription rates. Combined subscriptions for print copy plus access to the electronic version of journals typically are more expensive than print-only subscriptions. While the reasons for price increases may be understood by librarians, it can be very difficult to explain them to those who decide on the size of library budgets. As a consequence, we have to fight even harder for our budgets now.

3.2. PROCESSING ELECTRONIC PUBLICATIONS IN THE LIBRARY

Boyce and Dalterio (1996) pointed out that electronic publishing means different things to different people. With regard to the classic publication model author–publisher–library–reader, they identified six phases of electronic publishing: author preparation and submission of the electronic manuscript, peer review, copy editing and typesetting, database preparation, production and distribution, and archiving. To librarians, access to and distribution of electronic publications as well as archiving (which will be dealt with in Section 3.5.) are of obvious importance.

While book-like electronic media like CD-ROMs and diskettes are processed in libraries very similarly to print media, remotely stored electronic publications bring upon us changes that effect all phases of collection development. Selection criteria for purchasing material probably remain more or less as before (Nisonger 1996), but it might be more difficult to obtain evaluative information (e.g. book reviews) about electronic material. In January 1997, the first journal devoted to providing such evaluative information was launched[1] and may find a lot of approval among acquisitions librarians.

Once the decision to purchase (or lease) electronic publications has been made, there are often technological implications in trying to access the documents. Special software might be necessary that needs to be properly installed on local PCs or made available through the network. The library needs to provide public terminals and printers to ensure that patrons visiting the library can access the material. Because of the extensive use of the World Wide Web in the scientific community, it is hoped that a standardized way of providing electronic documents (i.e. in SGML [Standard Generalized Markup Language] or HTML [Hypertext Markup Language] format) will evolve that will overcome the various and confusingly different software packages currently used by publishers.

Cataloging needs to include all the bibliographic details catalog records for paper-based documents contain, plus information about access procedures, electronic format, size of files, hardware and software requirements and other technical details. Most libraries that have automated their catalog use the MARC (MAchine Readable Cataloging) formats for bibliographic representation. MARC currently is being enhanced by a subset of fields that cater to additional descriptions of electronic publications, including the electronic address of the document. Library catalogs with a hyperlinked World Wide Web interface can then provide access to electronic publications from within the catalog just by clicking on the address.

Access control can be taken care of by the publisher through IP (Internet Protocol) addresses, or it can be left as the responsibility of the subscriber (i.e. the library) to inform authorized readers about the user ID and password. The latter solution is not preferred by librarians, as it is nearly impossible to distribute a password to the library's user community and at the same time avoid any possible misuse by persons not entitled to access.

Costs of electronic resources will be difficult to calculate from now on. Publishers have come up with a large variety of new pricing strategies (see Brown & Duda 1996); "pay per view" and "flat fees plus additional charges according to usage" are just two examples. Pricing models are still in flux and changes to current solutions are inevitable. In addition, different libraries may have to pay different prices for journal subscriptions. Some publishers have started to offer special conditions and reduced prices to consortia of libraries. Participants need to pay only a portion of the total subscription costs, but gain access to all the publisher's journals. For

---

[1]Electronic Resources Review, http://www.mcb.co.uk/liblink/err/jourhome.htm

large university library systems, this model might be advantageous, but for small individual libraries it can be difficult or impossible to participate in a consortium for administrative, organisational or political reasons.

Instead of owning the material, there is a clear shift towards providing access to electronic documents (electronic journals in particular) just for a given time. Libraries are leasing electronic material rather than purchasing it. The necessary contracts, so-called license agreements, raise another set of questions (see e.g. Soete 1997): What exactly are the rights and responsibilities of the parties of the contract? Are there limitations regarding simultaneous accesses, downloading or printing of material? What are the technical requirements? Is the library entitled to access the user statistics in order to evaluate usage or does the publisher keep these data secret?

One of the most critical parts of license agreements concerns copyright. Publishers from many countries currently allow libraries to take photocopies of their print publications based on "fair use", i.e. if it is for personal, not commercial, use. This naturally includes interlibrary loan, but ILL can be restricted or even prohibited by license agreements. A typical license currently favors the information provider over the information user (Okerson 1996a). If the restrictive clauses are not eliminated from the contracts, they will disable traditional and necessary cooperation among libraries. A positive exception is the text of the American Astronomical Society (AAS) contract. The AAS has included a librarian in their Publications Board for several years and the license agreement they have set up for their electronic journals is exemplary in its fairness and thoughtfulness[2]. The problem of negotiating contracts for electronic documents has led to extensive discussion among librarians. Documents that guide librarians through the jungle of licensing, for example the "Liblicense" resource[3] are necessary and very welcome. A newly established mailing list devoted to the topic of electronic content licensing, liblicense-l[4], received some 800 subscriptions within only a few days of its inception which clearly shows that discussion and advice on this topic are sought by many librarians.

It is the nature of leased material that it remains physically with the provider after the contract has ceased. Librarians must be aware that they may be left without anything when a license agreement terminates. This situation is unlike the print environment where a library at least owns the

---

[2]The Astrophysical Journal and Astrophysical Journal Supplement Series Institutional Site License, http://www.journals.uchicago.edu/ApJ/sitelicense.html

[3]Liblicense: Licensing digital information. Yale University Library, http://www.library.yale.edu/~Llicense/index.shtml

[4]To subscribe send a message to listproc@pantheon.yale.edu, leave the subject line blank and type in the body of the message: subscribe LIBLICENSE-L Firstname Lastname

journal volumes that were published while a subscription lasted. Some journal publishers deliver backups of electronic journals on CD-ROMs which stay in the library even after cancellation of the subscription. In addition to the problem of whether or not CD-ROMs will be a viable long-term storage medium, they are not capable of presenting one of the major advantages of electronic over print journals, i.e. they are not interlinked with other journals or databases. Backup on CD-ROM therefore is not an appropriate alternative. The only fair solution would be to allow libraries access to those volumes they subscribed to.

3.3. ELECTRONIC BOOKS AND NON-SERIAL DOCUMENTS

Everybody who enjoys the calming sound of turning book pages most probably dislikes the thought of having to do with a computer mouseclick instead. Today's technology is advanced, but not yet advanced enough for most of us to really want to read electronic books. The general market seems to reflect this, since it has not developed as steadily as it was predicted only a few years ago. The 1996 Frankfurt Book Fair, one of the world's largest international book fairs, saw a smaller percentage of multimedia products than anticipated (Baier 1996). Klaus-Dieter Lehmann, Director of the Deutsche Bibliothek stated in an interview that only 2,000 to 3,000 out of 300,000 new acquisitions per year are digital, everything else still is on paper (Lehmann 1996).

Electronic books do have some advantages though. They allow users to quickly locate specific information through built-in search functions, and information can be updated and kept accurate easily. Electronic format therefore is a valid alternative to print media, especially with regard to directories, dictionaries, encyclopedias, and reference works.

In astronomy, extensive use of the capabilities of e-books has been made in regard to user manuals for telescopes or individual instruments. These days the latest information often can only be found on the Web. For conference announcements and programs, it has become standard for users to obtain up-to-date information from a Web page. Likewise, more and more conference proceedings can be found on the Internet (e.g. "Weaving the Astronomy Web [WAW]"[5], "Library and Information Services in Astronomy II [LISA II]"[6] or the "Astronomical Data Analysis Software and Systems [ADASS]"[7] conference proceedings). A novelty in astronomy was the book "Information & on-line data in astronomy" (Egret & Albrecht 1995); in ad-

[5]WAW, http://cdsweb.u-strasbg.fr/waw/proceedings.html
[6]LISA II, http://www.eso.org/lisa-ii/lisa-ii.html
[7]ADASS, http://iraf.noao.edu/ADASS/adass.html

dition to the paper version, an electronic complement is maintained on the World Wide Web, from where chapter abstracts are available. Hypertext links mentioned in the book are provided as clickable links.

On the other hand, electronic books and documents have specific disadvantages. With regard to networked documents, version control poses a distinct problem. We are used to informing our users about the edition of a book or the publication year of a thesis, and we know that the document has not been changed since it arrived on paper in our library. In contrast to this, information about publication dates and versions of documents that are accessed over networks are extremely short-lived. The responsibility for these documents remains with the author or publisher and thus they can be changed by them, if deemed appropriate. The dynamic nature of networked documents leads to the dilemma of up-to-date information versus access to the original version. Von Ungern-Sternberg and Lindquist (1995) pointed out that another concern is the authenticity or integrity of electronic documents. Once a document is stored on a computer, it can be relatively easily manipulated by additions, deletions or changes. "A number of 'digital signatures' or 'electronic seals' based on cryptographic techniques" may help to ensure authenticity. If electronic books are delivered offline (e.g. on diskettes, CD-ROMs or their successors), the difficulties of version control and validation of authenticity diminish. However, offline publications cannot be updated as easily as networked documents, and they don't exploit the capability of interlinking with external resources. Instead, hypertext links are used to navigate within the document or in order to make use of built-in multimedia effects like sound, video, and simulations.

Although an unlimited number of copies can be downloaded or printed from electronic books, assuming the electronic version is free-of-charge or the access fee has been paid, nevertheless electronic books can become unavailable. Out-of-print material on paper almost always can be located in some way, most importantly through interlibrary loan, and therefore hardly ever disappears entirely. Online documents, in contrast, might be withdrawn by the authors or publishers and may never be found again.

On the Internet, it is astonishingly easy to become a "publisher". Everybody equipped with a networked terminal and a server has the opportunity to distribute information globally without any difficulty. While this can be seen as moving towards more democracy in a general context, it is a threat in scientific publishing, because quality standards are likely to decrease. Algorithms that try to match users' requests with available documents do not solve this problem, because "prioritizing according to relevance won't give you a true measure of quality" (Schwartz 1996). Commercial publishers and learned societies therefore argue that a careful editing process is necessary to assure that only those documents which would pass the quality control for paper-based publication are published electronically. Unfortunately, such editing is one of the major cost factors of journal production which will not disappear even if the journal is available exclusively in electronic format.

3.4. ELECTRONIC JOURNALS

We are still in the early days of electronic journals. This phase brings upon us a variety of formats and delivery methods. Formats include bitmaps, PostScript, PDF (Portable Document Format) as well as ASCII (American Standard Code for Information Interchange), SGML and HTML. Electronic journals might be delivered on CD-ROM, by electronic mail, or through the networks. Networked journals require that users actively access the publishers' sites; for convenience of the subscribers, new issues often are announced through e-mail notifications. Although publishers are still experimenting with various formats, the extensive use of the World Wide Web certainly will lead to a future of electronic journals that is network-based (Wusteman 1996).

The increase in the number of so-called electronic journals is breathtaking. In the introduction to the 1996 edition of the "Directory of Electronic Journals, Newsletters and Academic Discussion Lists", Ann Okerson noted that "in this sixth edition, the number of journal and newsletter titles (nearly 1700) has more than doubled since last year's and multiplied by over 15 times since the first edition (there were 110 listings in 1991, 240 in 1993, and nearly 700 last year)" (Okerson 1996b). Most of these journals are not "true electronic journals" though. In a widely distributed survey article by Steve Hitchcock, Leslie Carr and Wendy Hall (1996), the authors point out that currently most of the electronic journals are rather electronic versions of paper journals than fully operational e-journals; that is, they are confined to presenting bitmapped images or electronic formats intended for printing, like PostScript or PDF. In order to fully exploit the capabilities and advantages of electronic journals, they must be able to interlink sources and therefore need to go beyond print-oriented format as soon as possible.

The acceptance of electronic journals among the scientific community currently is not as high as the large number of electronic versions seems to indicate. According to Ann Schaffner (1994), in order for e-journals to be accepted by authors, it is necessary that they fulfill at least the basic functions of paper-based journals, which she defines as

- Building a collective knowledge base
- Communicating information

- Validating the quality of research
- Distributing rewards
- Building scientific communities

It is obvious that these criteria are met by electronic versions of established print journals, but print-oriented formats only will be helpful in achieving acceptance during the transition phase from paper to electronic. Readers of a (paper) journal might accept more easily the electronic edition if it has the same "look and feel". Currently many scientists rely on paper, at least for reading full text articles. Interviews with chemists at Cornell University in 1995 showed that the possibility of creating a print copy determined to a large extent the scientists' acceptance of electronic journals (Stewart 1996). However, in the long run "authors are unlikely to accept 'electronic clones' of paper journals" (Ginsparg 1996).

Accessing electronic publications over the Internet can be unacceptably slow, and incorrect or out-of-date electronic addresses can turn locating documents into an extremely time-consuming task. Reliable links are the backbone of electronic services. A study by Harter and Kim (1996) revealed that 55% (71 out of 129) of the links to e-journal archives the authors tried to access did not work at first try. If electronic journals are to be effectively used, the reliability of electronic references must stabilize.

Harter and Kim also analyzed to what extent authors of e-journal articles (in whom one can assume a sympathetic attitude towards electronic media) are citing electronic journals and other online resources (electronic personal papers, newsgroup postings, electronic preprints etc.). Their findings showed that only 1.9% (83 out of a total of 4,317) of the citations in peer-reviewed articles published in 1995 referred to electronic sources. Indexing services can help increasing the acceptance of electronic publications but they are just beginning to include e-journals in their databases. Electronic journals must struggle to meet the criteria for inclusion applied by abstract services in order to be eligible.

One of the most important problems of electronic publications is archiving. Unless a stable system for future retrieval of publications is in place, authors will continue to be reluctant to publish in electronic journals.

3.5. ARCHIVING

In order to guarantee continued access to information, publications have been archived throughout time. Traditionally, preservation has been a task of libraries, which they have performed reliably through the centuries. Up to now, even small specialized libraries were able to provide a highly valuable repository of contemporary and historical publications. Their holdings could be unique within their geographical region.

We have seen many media tested for and applied to archiving, for example paper, microfilm, microfiche, CD-ROM, and magnetic tape. Of all these, paper still is the preferred storage medium for many librarians, since it is the only format that can be used without any special viewing devices and therefore is independent of current and future technology. In addition, no other storage medium up to now has survived for as long as paper.

Electronic publications turn this ordered situation upside-down. They unfold their most valuable features when they are interlinked with other documents, archives, and databases. Thus, if we regard electronic publications not as self-contained, off-line documents, but as parts of a large system, we must conclude that they need an archiving system that is based on a global concept and goes beyond our current model.

A new physical storage medium (e.g. magnetic tape, networked on computers) as well as an appropriate data format (e.g. SGML) have to be selected (Wusteman 1996). It is obvious that current technology soon will be replaced by newer versions or entirely new systems, therefore both the physical storage medium and the data format must be sufficiently flexible to be transferred to the next technological generation. This transfer is being referred to as "technology refreshing" or, as an even broader concept, "technology migration" (Garrett & Waters 1996).

The rapid pace at which technology becomes obsolete makes archiving an extremely expensive undertaking. Libraries' budgets are highly unlikely to rise with the necessary speed (in fact, they are unlikely to rise at all). Librarians may find themselves being forced to abandon archiving, one of their traditional tasks. New volunteers appear on the scene, some eager to take over. In addition to non-profit organizations, commercial institutions and subscription agencies participate in the debate. Many librarians are skeptical about whether the interest of commercial organizations in archiving will go beyond immediate economical consideration. We also wonder whether we will have to pay twice for journals in the end – once for the license, and again for access to back issues.

In contrast to commerce-oriented solutions, the Task Force on Archiving of Digital Information, established by the Commission on Preservation and Access and the Research Libraries Group, suggests in its final report that a national system of digital archives should be developed (Garrett & Waters 1996). Digital archives in their sense are "held together in a national archival system through the operation of two essential mechanisms": the repositories must meet or exceed criteria set by a program for archival certification, and a "critical fail-safe mechanism" must be in place that can carry out rescue functions when needed. This report focuses on the United States and although such a well-organized, nation-wide archival system will be difficult, maybe even impossible, to achieve for many countries, the Task Force envisions a possible expansion on an international level.

International standards for archiving electronic publications in conjunction with a stable and mature network environment would be the ideal solution, but it will take some time before the necessary infrastructure will be in place. In the meantime, we must find other solutions (see e.g. Neavill & Sheblé 1995). We have reached a critical point in time, because technology has become so advanced that publishers are tempted to abandon the traditional archiving medium before actually being able to come up with another long-term solution that is appropriate for the digital age. If the mere availability of technology leads to short-sighted solutions, it will have expensive, if not fatal implications for future access to scientific literature. In this regard, the opportunities given to us by technology clearly are a danger. In order to ensure uninterrupted access to publications, archiving must continue on paper until the question of infinite preservation is fully clarified.

4. Indexing and Retrieving Electronic Publications

Locating information on the Internet seems to be so easy. Every publication on the network has its proper address, and all we need to do in order to retrieve it is to point our Web browser to that address.

From their daily work, librarians know that the reality is different. In paper-based publications, we have come across many strange or wrong citations over the course of the years. If this is the case in the relatively standardized and agreed-upon print environment, it is unlikely that this situation will improve when it comes to electronic publications. The current naming system for networked documents relies on a chain of words, letters, or figures, subdivided by commas, dots, slashes, hyphens or other characters. The so-called Uniform Resource Locators (URL) are extremely liable to errors. One lower-case letter instead of a capital one will result in an error message.

Librarians and publishers are experimenting with better naming systems. Uniform Resource Locators are already developing into Uniform Resource Names (URN), which allow a unique name to be assigned to the publication that will not be changed even if the document is moved to another computer. A name resolver will keep track of the actual location of the requested file and translate the name into the correct URL. The American Astronomical Society makes use of this approach in their electronic journals (Warnock & Fullton 1996). OCLC (Online Computer Library Center,

Inc.) has been testing Persistent Uniform Resource Locators (PURLs)[8] that also point to an intermediate resolution service instead of directly to the location of an Internet resource.

These mechanisms are not yet commonly used, and librarians have to be imaginative in order to locate requested documents when users present to them strange network addresses and weird names of electronic publications. Our profession demands that we always try more than just one approach in order to locate information. If the most obvious way does not lead to satisfactory results, we think of alternatives. This attitude is being extended to Internet resources where usually several search strategies are possible.

For many years, librarians have been the experts within their institutes with regard to searching commercial databases. They often even held a monopoly, because they were the only ones who knew which retrieval language had to be used for which database and how required information could be obtained most cost-effectively. It is obvious that searching the seemingly chaotic Internet is completely different from well-organized bibliographic or full-text databases and therefore requires new methods. The most commonly used resource discovery tools are search engines that allow searching the Internet for words or phrases. The drawback is that enquiries typically result in a huge number of documents with extremely high noise and little precision. Projects have been set up in order to improve search results. What is needed is information about information, so-called metadata. OCLC and NCSA (National Center for Supercomputing Applications) are two of the main initiaters who proposed the Dublin Core Metadata elements set which is "intended to describe the essential features of electronic documents that support resource discovery" (Weibel *et al.* 1996). The 13 Dublin Core elements include, for instance, information about the subject, title, authors, form, etc. and can be generated by authors of Web documents without extensive training. "Indexing the Internet" sounds like a mission impossible but the achievements so far are very promising. If some remaining problems can be solved, it is quite possible that the Dublin Core will be implemented in one of the future versions of HTML. Mapping between MARC and the Dublin Core also is in preparation so that data interchange between catalog records and SGML documents will be possible (Dempsey & Weibel 1996 and Weibel 1995)[9].

This approach to information retrieval still involves human mediation. In a recent article, Bruce Schatz (1997) explains how new concepts may

[8]PURL, http://purl.oclc.org/

[9]For more information on metadata see also: IFLA, Digital Libraries: Metadata Resources, http://www.nlc-bnc.ca/ifla/II/metadata.htm and Library of Congress: Metadata, Dublin Core and USMARC: a review of current efforts, Discussion Paper No. 99, gopher://marvel.loc.gov/00/.listarch/usmarc/dp99.doc

allow users to search through distributed repositories across the net. He foresees that automatic indexing and "vocabulary switching" (i.e. terminology being automatically "translated" into the appropriate vocabulary of different subject domains) finally will allow us to "effectively utilize the whole of scientific information." Although such systems are still under development, one of the first applications can be seen already in the Illinois Digital Library Initiative[10] in which the AAS is also about to participate.

5. The Role of Librarians in the Electronic Environment

The mission of astronomy librarians typically is to provide access to selected information resources according to the needs of our users, to collect and preserve material, to organise it, and to deliver it to the requester. This mission statement refers to all available media and document formats and therefore is not limited to particular technologies or specific software applications. Librarians are system-independent information specialists, aiming at fulfilling the information needs of library users. Our experience indicates that user needs and expectations are very similar in the paper-based and in the electronic environment:

> Users' needs will continue to be what they long have been. Users will want information reliably locatable [...] (and) easily accessible [...]. In the electronic environment the need for access tools will be more evident [...]. Users will expect information to be available that was placed in the library's care a long time ago; and they will expect that the integrity of the information they get from the library will be assured. (Graham 1995)

As in the paper-based environment, digital services must be "planned, implemented, and supported" (Hastings & Tennant 1996), and instead of being the "gatekeepers" to material, librarians now follow new directions which were identified by Ching-chih Chen (1994) as follows:

- from library-centered to information-centered
- from the library as an institution to the library as an information provider, and the librarians as a skilled information specialist functioning in an all-related information environment
- from using new technology for the automation of library functions to utilizing technology for the enhancement of information access and delivery of items not physically contained within the four walls of the library

---

[10]Digital Library Initiative project at the University of Illinois at Urbana-Champaign, http://dli.grainger.uiuc.edu/

– from library networking for information provision to area networking
  for all types of information source providers

Chen continues to add another important aspect, namely "from access
to selectivity". It is not lack of information we have to cope with, but
an information overload that needs to be mastered. "End-user searching"
is a frequently used term today, since (at least seemingly) user-friendly
retrieval technologies are evolving. However, recent studies revealed that
"the preferred method of receiving information by many of our clientele
is person to person" and that "service is personal" (see Enyart & Smith
1996). A report about the University of California Digital Library draws
the same conclusion: "Experience indicates that, despite the availability
of intelligent systems, increasing remote access will also increase demands
for service, both online and face-to-face" (University of California Library
Council 1996).

Since the beginning of our profession, librarians have been in direct contact with users seeking information. We have learned to understand what
they need, not what they say they need (which can be considerably different). Our services are personalized and targeted for our clientele, and up to
now no "interface agents" and "personal filters" as described by Nicholas
Negroponte (1996, pp. 149-159) are in place that are able to substitute
person-to-person mediation.

Technology for producing and distributing information is useless without some way to locate, filter, organize and summarize it. A new profession of "information managers" will have to combine the skills of
computer scientists, librarians, publishers and database experts to help
us discover and manage information. These human agents will work with
software agents that specialize in manipulating information – offspring
of indexing programs such as Archie, Veronica and various "World Wide
Web crawlers" that aid Internet navigators today. (Varian 1995)

What Hal Varian overlooks is the fact that this profession exists already. The "human agents" he mentions are the librarians of today, no
matter whether they call us information navigators, information officers,
cybrarians, or librarians, a term most of us still prefer (Ojala 1993).

In the electronic environment, our role comprises at least three major
working areas: We offer services for those users who want to be guided to
the most suitable information resources; we provide research assistance for
those who prefer to conduct searches themselves and only turn to librarians
in case their repertoire of search and retrieval techniques did not lead to
satisfying results; and we closely collaborate with Information Technology
(IT) departments that design new and enhance existing systems. These
working areas encompass the following aspects:

1. Information Access Provider:
   We provide access to the most important information resources on the network, making use of current technologies (WWW or its successor). Electronic resources must be organized in a logical, easily understandable manner, integrating documents and services that belong logically together. The purpose of value-added services like subject-oriented clearinghouses is "not only to save the researcher time and effort in searching for appropriate sources in the vastly unordered, unstructured Internet, but also to provide him or her with a pre-assessed, semi-ordered, annotated list of sites with activatable links" (Rusch-Feja 1997) which match or supplement the targeted group's information needs.

2. Research Assistant:
   The second aspect of our role encompasses identifying, locating and obtaining publications not owned or leased by the library. No resource, be it electronic or on paper, can be called "complete" today, and if we don't find particular information on the Internet, this does not mean that it doesn't exist. It is the librarian's duty to know which additional sources can be queried to make a search as complete as possible. Publications "not imbedded in a formal journal context" (Rusch-Feja 1997) as well as electronic equivalents to today's "grey literature" need to be retrieved, which requires the librarian's experience in locating information as well as technological skills. Research assistance in this sense also includes helping our users to become familiar with handling new technologies. Various methods for providing user support for networked library services can be developed, for instance distance support (by telephone or e-mail), printed or online manuals, and on-screen instructions (Mackenzie Owen & Wiercx 1996). Face-to-face end-user training already has become an important part of our work.

3. Collaborative System Designer:
   Librarians are in direct contact with users of information retrieval systems. Often users report difficulties to us which they encountered while using a system, or we recognize what needs to be redesigned while we are explaining a system to our users. When IT departments design new databases, application programs and user interfaces for us, we must be able to explain precisely to them "how language works and how to use layout, typography and design principles" (Moore 1996) in order to provide the required functionality. The information flood can only be mastered with appropriate tools that are capable of matching user needs with the available information, no matter whether these tools

will be operated by librarians or users in the end.

Access to global networks and the availability of information to scientists through these networks have changed vastly how science is done today. Despite the many advantages of electronic publishing, there are still many problems to be solved. The information overload results in a need for specialists who are experienced in retrieving information and therefore can guide end-users through the jungle of available resources and technologies. In the Information Age, professional librarians are likely to be more in demand than ever.

6. Conclusion

As we are heading towards the 21st century, we are witnessing an information revolution. The importance and value of information, delivered in a timely manner to the requester, is steadily growing. New technologies have changed dramatically the ways in which information can be obtained.

Astronomy librarians are in the fortunate situation of having tools in their hands that provide search and navigation capabilities of which we wouldn't have dreamed only a few years ago. On the other hand, the current situation also confronts us with a large number of questions. Librarians are not interested exclusively in accessing the very latest (pre-)publication, but need to make sure that scientific documents will be available throughout time. Many librarians are alarmed by some of the directions currently suggested. Mature and standardized solutions for retrieving and accessing and in particular for archiving electronic publications must be in place before the current, paper-based system can be abandoned.

Librarians traditionally have been on the forefront of making information available to users. They will continue to be so, but the means and methods used in order to achieve this aim are undergoing rapid changes. The role of librarians in the new publishing scenario has been expanded, it involves new tools and technologies, and reflects changed requirements and new end-user behavior.

Acknowledgements

I am very thankful to Ellen Bouton (National Radio Astronomy Observatory, Charlottesville), Marlene Cummins (University of Toronto Astronomy Library, Toronto), and Sarah Stevens-Rayburn (Space Telescope Science Institute, Baltimore) who provided a world of help through thoughtful comments and suggestions, as well as through careful proofreading (that was how I learned that the four of us can be called "fuss-buckets"). Special thanks go to Carolina Noreña Janssen (Museo de Ciencias Naturales, Madrid) for extensive discussions and helpful comments.

References

1. Baier, H. (1996) Frankfurter Buchmesse: Seismograph für Trends, *Buch und Bibliothek* **48-12**, 895-896
2. Boyce, P.B. & Dalterio, H. (1996) Electronic publishing of scientific journals, *Physics Today* **49-1**, 42-47
3. Boyce, P.B. (1996) Building a peer-reviewed scientific journal on the Internet, *Computers in Physics* **10-3**, 216-221
4. Brown, E.W. & Duda, A.L. (1996) Electronic publishing programs in science and technology. Part 1: The journals, *Issues in Science and Technology Librarianship* **Fall**, http://www.library.ucsb.edu/istl/96-fall/brown-duda.html
5. Butler, D. (1996) Internet growth – boom or bust? *Nature* **380**, 4 April, 378
6. Chen, C. (1994) Information SuperHighway and the Digital Global Library: realities and challenges, *Microcomputers for Information Management* **11-3**, 143-155
7. Corbin, B.G. & Coletti, D.J. (1995) Digitization of historical astronomical literature, *Vistas in Astronomy* **39-2**, 127-135. Electronic version at http://www.eso.org/lisa-ii/lisaii-abstracts.html#abs11
8. Dempsey, L. & Weibel, S.L. (1996) The Warwick Metadata Workshop: a framework for the deployment of resource description, *D-Lib Magazine* **July/August**, http://www.dlib.org/dlib/july96/07weibel.html
9. Egret, D. & Albrecht, M.A. (eds.) (1995) *Information & on-line data in astronomy*. Kluwer Academic Publishers, Dordrecht. Electronic complement at http://cdsweb.u-strasbg.fr/data-online.html
10. Enyart, M.G. & Smith, R.A. (1996) Reference services: more than information chauffeuring, *Special Libraries* **87-3**, 156-162
11. Garrett, J. & Waters, D. (co-chairs) (1996) *Preserving digital information*. Report of the Task Force on Archiving of Digital Information commissioned by The Commission on Preservation and Access and The Research Libraries Group. http://www.rlg.org/ArchTF/tfadi.index.htm
12. Ginsparg, P. (1996) Winners and losers in the global research village, *Joint ICSU Press-UNESCO Expert Conference on Electronic Publishing in Science*, Paris, France, February 19-23, 1996. http://xxx.lanl.gov/blurb/pg96unesco.html
13. Graham, P.S. (1995) The digital research library: tasks and commitments. In: *Digital Libraries '95*, Austin, Texas, 11-12 June 1995. Electronic version at http://csdl.tamu.edu/DL95/papers/graham/graham.html
14. Harter, S.P. & Kim, H.J. (1996) Accessing electronic journals and other e-publications: an empirical study, *College & Research Libraries* **57-5**, 440-456
15. Hastings, K. & Tennant, R. (1996) How to build a digital librarian, *D-Lib Magazine* **November**, http://www.dlib.org/dlib/november96/ucb/11hastings.html
16. Hitchcock, S., Carr, L. & Hall, W. (1996) A survey of STM online journals 1990-95: the calm before the storm. In: *Directory of Electronic Journals, Newsletters and Academic Discussion Lists*, D.W. Mogge (ed.), 6th ed., Association of Research Libraries, Washington, D.C., pp. 7-32. Electronic version at http://journals.ecs.soton.ac.uk/survey/survey.html
17. Lehmann, K.-D. (1996) Bücher haben Vorteile, *Der Spiegel* **51**, 184-189
18. Mackenzie Owen, J.S. & Wiercx, A. (1996) *Knowledge models for networked library services – EUR 16905*. Office for Official Publications of the European Communities, Luxembourg. (Libraries in the information society series)
19. Michael, J. (1994) Responding to the revolution: becoming the library without walls. In: *From A to Z39.50: A networking primer.* J.J. Michael & M. Hinnebusch. Meck-

lermedia, London, pp. 32-38

20. Moore, N. (1996) Creators, communicators and consolidators: the new information professional, *Managing Information* **3-6**, 24-25

21. Neavill, G.B. & Sheblé, M.A. (1995) Archiving electronic journals, *Serials Review* **21-4**, 13-21

22. Negroponte, N. (1996) *Being digital*. Vintage Books, New York.

23. Nisonger, T.E. (1996) Collection management issues for electronic journals, *IFLA Journal* **22-3**, 233-239

24. Ojala, M. (1993) What will they call us in the future? *Special Libraries* **84-4**, 226-229

25. Okerson, A. (1996a) Who owns digital works? *Scientific American* **275**, July, 64-68

26. Okerson, A.S. (1996b) Introduction. In: *Directory of Electronic Journals, Newsletters and Academic Discussion Lists*, D.W. Mogge (ed.), 6th ed., Association of Research Libraries, Washington, D.C., http://poe.acc.virginia.edu/~pm9k/libsci/96/intro.html

27. Postman, N. (1992) *Technopoly: the surrender of culture to technology*. A.A. Knopf, New York.

28. Rusch-Feja, D. (1997) Subject-oriented collection of information resources from the Internet: a clearinghouse concept to support scientists in a German research institute, *Libri*, in press

29. Schaffner, A.C. (1994) The future of scientific journals: lessons from the past, *Information Technology and Libraries* **13**, 239-247

30. Schatz, B.R. (1997) Information retrieval in digital libraries: bringing search to the net, *Science* **275**, 17 January, 327-334

31. Schwartz, R.A. (1996) A sea change for academic publishing, *ASEE Prism* **5-9**, 12-16

32. Soete, G. (1997) Let there be light! Licensing electronic resources: state of the evolving art. Summary of proceedings, San Francisco, USA, December 8-9, 1996. Version as of January 6, 1997. http://arl.cni.org/scomm/sum.html

33. Stewart, L. (1996) User acceptance of electronic journals: interviews with chemists at Cornell University, *College & Research Libraries* **57-4**, 339-349

34. University of California Library Council (1996) The University of California Digital Library: a framework for planning and strategic initiatives. Version as of October 31, 1996, updated November 19, 1996. http://sunsite.berkeley.edu/UCDL/title.html

35. Varian, H.R. (1995) The information economy. How much will two bits be worth in the digital marketplace? *Scientific American* **273**, September, 161-162

36. Von Ungern-Sternberg, S. and Lindquist, M.G. (1995) The impact of electronic journals on library functions, *Journal of Information Science* **21-5**, 396-401

37. Warnock, A. and Fullton, J.M. (1996) The Electronic Astrophysical Journal: resource location and archive management. In: *Astronomical Data Analysis Software and Systems V*, G.H. Jacoby and J. Barnes (eds.), ASP Conference Series **101**, 589-592

38. Weibel, S.L. (1995) The World Wide Web and emerging Internet resource discovery standards for scholarly literature, *Library Trends* **43**, 627-644

39. Weibel, S., Godby, J., Miller, E. and Daniel, R. (1996) OCLC/NCSA metadata work-shop report. http://www.oclc.org:5046/conferences/metadata/dublin_core_report.html

40. Wusteman, J. (1996) Electronic journal formats, *Program* **30-4**, 319-343

THE ROLE OF DATA CENTRES
IN THE ERA OF ELECTRONIC PUBLISHING

How CDS weaves links towards astronomical databases and archives

D. EGRET AND F. GENOVA
Centre de Données astronomiques de Strasbourg,
Observatoire Astronomique
11 rue de l'Université
F-67000 Strasbourg, France
egret@astro.u-strasbg.fr
http://cdsweb.u-strasbg.fr/people/de.html
genova@astro.u-strasbg.fr
http://cdsweb.u-strasbg.fr/people/fg.html

Abstract. The astronomy data centres, and in particular the *Centre de Données astronomiques de Strasbourg* (CDS), have been building electronic information services for many years. References of publications, observational data related to objects, data tables, nomenclature, have been homogenized and organized into information retrieval systems.

This undertaking implied an effort of collaboration between data centres, data providers, agencies, journal editors, etc. Evolution in recent years has brought the data centres closer from the publishing process. General standards for electronic tables, tabular data, and catalogues have been proposed and implemented.

With the emergence of fully electronic publication, new digital library services are being organized, and pave the way to innovative new services, linking publications to information from other sources, and making use of new methods for textual information retrieval. The data centres expect to play a key rôle in these new developments, taking advantage of their expertise in the development of value-added services, and of their long-term involvement towards a fully linked astronomy information system.

1. Introduction

In the present context of an unprecedented accumulation of data collected

Astrophysics and Space Science **247**: 175–188, 1997.
© 1997 *Kluwer Academic Publishers.*

by ground or space-based telescopes and observatories, the rôle of the data centres is to bridge the gap between the specialized approach of the scientific teams (where the detailed expertise about the specific data resides), and the general approach of the wider community of researchers (who need an easy access to data, calibrated as far as possible in meaningful physical units).

The concept of data centre had to be somehow extended to the one of 'information hub' where the users find an organized set of services and tools to help them retrieving the data and information they need among the many possible distributed on-line resources.

The Strasbourg astronomical Data Centre (CDS), hosted by the Strasbourg astronomical Observatory, and funded by the French *Institut National des Sciences de l'Univers* (INSU, CNRS) is the prototype of such an astronomical information hub (Genova *et al.* 1996).

We present here the current efforts of CDS for providing on-line data and information for the world-wide astronomy community, in the specific context of the coming era of electronic publishing and of increasingly integrated digital library services.

Then, we discuss the new perspectives of a global digital library service and show the new rôle we expect to be played by data centres.

2. The CDS services

The objective of the Strasbourg astronomical Data Centre (CDS) is to provide on-line data and information for the world-wide astronomical community. This is done through development of a set of complementary information services (see e.g. Egret *et al.* 1995a & 1995b): we will briefly describe here these services, giving special attention to the possible interactions with electronic publishing.

The common gateway to most of these services is the CDS home page on the World-Wide Web (http://cdsweb.u-strasbg.fr/CDS.html).

2.1. THE SIMBAD DATABASE OF ASTRONOMICAL OBJECTS

2.1.1. *SIMBAD main features*
The specificity of the SIMBAD database is to organize the information per astronomical object, thus offering a unique perspective on astronomical data. This can only be done through a careful cross-identification of objects from catalogues, lists, and journal articles. The ability to gather together any sort of published observational data related to stars or galaxies has made SIMBAD a key tool used worldwide for all kinds of astronomical studies.

The SIMBAD database has been described by Egret *et al.* (1991). We will just recall here, for completeness, the main features of this astronomical object-oriented database:

- a database of more than 1,400,000 astronomical objects (stars, galaxies and all astronomical objects outside the solar system);
- a cross-index to more than 2200 astronomical catalogues, lists, and observation logs of space missions;
- observational data from some 25 different types of data catalogues and compilations;
- a bibliographic survey covering the astronomical literature since 1950 for stars, and since 1983 for extragalactic objects;
- a management of the possible variations in the naming of astronomical objects with the *sesame* module and the Dictionary of Nomenclature of Celestial Objects (see below);
- a *name resolver* integrated within the archive systems of major observatories (Hubble Space Telescope, European Southern Observatory, Canada–France–Hawaii–Telescope, etc.), and which is also a powerful tool for bibliographical services.

SIMBAD is kept up-to-date on a daily basis, as the result of the collaboration of CDS with bibliographers in Institut d'Astrophysique de Paris and the Paris and Bordeaux observatories (Laloë *et al.* 1993 and Laloë 1995) who systematically scan the articles published in more than 90 astronomy journals.

Large astronomical catalogues, carrying their specific identifiers and measurements, are added after a cross-matching procedure which may span over many months, or even years.

The data contained in SIMBAD are also permanently updated, as a result of errata, remarks from the librarians (during the scanning of the literature), quality controls, or special efforts from the CDS team to better cover some specific domains. Requests for corrections, errata, or suggestions are regularly received from SIMBAD users through a dedicated *hot line*, at e-mail address `question@simbad.u-strasbg.fr`.

There are at least three key aspects which make SIMBAD a specifically powerful tool in the era of electronic publishing:

- SIMBAD can be seen as a dictionary of identifications, aliases and names of objects: in principle any name found in the literature — and provided it is given as a syntactically correct character string— can be submitted to SIMBAD in order to retrieve basic information known for this object, as well as pointers to complementary data and bibliography.

- Another unique feature is the complete and up-to-date bibliographic survey of the astronomical literature: objects are associated with the references of all papers in which the objects is mentioned, under any of its aliases.
- In addition, the Dictionary of Nomenclature (see Section 2.1.3 below) is critical for managing the very complex nomenclature of objects found in the literature, and for matching these naming variations with those adopted or simply accepted by SIMBAD; it also includes hints for helping to solve ambiguities, according to the type of object, or to the format.

2.1.2. *The SIMBAD bibliography*

The SIMBAD bibliography for the astronomical objects includes references to all published papers from some 90 periodicals covering the whole astronomical literature. Articles are scanned in their entirety, and references to all objects mentioned are included in the bibliography. References, authors, and titles are stored for about 100,000 papers since 1950 providing some 2 million references to astronomical objects.

The scanning of the published literature is currently made manually, mainly by librarians under the responsability of Institut d'Astrophysique de Paris. The advent of electronic publication brings obviously new perspectives for improvement and automation of this procedure. In a first place, Table of Contents are now received electronically through the network, thanks to journal Editors, thus reducing the risk of errors. Regularly, a number of electronic lists of objects are also folded into Simbad through semi-automatic procedures. The next step will be the automatic flagging of object names in the text of the articles: this is not yet done, but is certainly a very interesting short-term goal.

Two ways of achieving this flagging are currently being considered:

- the first one is to ask the authors, with the help of the Editors of electronic journals, to flag astronomical object names in his/her text; this can be done, for instance, by the use of a \astrobj{ } command within the TeX or LaTeX source, which will be eventually used to build an anchor pointing towards SIMBAD, or a similar database, in the on-line version made available on the network.
- another approach is the use of intelligent search tools for identifying object names within the electronic version of the paper, using a set of syntactic and semantic rules, and the Dictionary of Nomenclature as a reference database for already known objects.

The first approach seems safer, provided the authors understand what exactly they are being required, and accept this (minor) additional work load. The latter implies a lot of fine tuning from the system developers. The current experience with the handling of publication suggests that both approaches may be needed, and that a careful quality control, including final check by an expert, will probably remain necessary to avoid errors or misinterpretations, and to ensure appropriate completeness.

2.1.3. *Dictionary of Nomenclature*

The question of nomenclature is becoming more and more crucial, as computer tools are being used, and object names are transmitted from one service to the other. The CDS, in collaboration with Observatoire de Paris (DASGAL), is now responsible of the development, maintenance, and distribution of the Reference Dictionary of the Nomenclature of Celestial Objects which has been originally published by Lortet *et al.* (1994).

The Nomenclature Dictionary can be queried through different keys (catalogue name, author, object type, format, etc.) as a World-Wide Web service, or directly within SIMBAD (where it is known as the `info` tool). Naming variations found in published papers are tracked in the Dictionary, with reference to the original list. The authors can also submit a new acronym, check that it is not already used, and have it approved by the Working Group on Nomenclature from Commission 5 of the International Astronomical Union (IAU).

Further developments will be to use directly the Dictionary as an auxiliary database in Simbad, for name resolving.

2.2. THE CATALOGUES OF OBSERVATIONAL DATA

2.2.1. *A depository of catalogues and archives*

The traditional rôle of CDS, as one of the first major astronomical data centres, is to collect catalogues, control data integrity, and act as a depository of catalogues and archives.

This activity is undertaken in the framework of international exchange agreements including NASA Astronomical Data Center (ADC), Japan's National Astronomical Observatory in Tokyo, the Russian Academy of Sciences, China's Beijing Observatory, and India's InterUniversity Center for Astronomy and Astrophysics.

2.2.2. *Electronic tables*

The traditional way to publish astronomical data (observational material, results from calibrations or models, etc.), was, until very recently, to submit the printed version to a refereed journal, and eventually make the corresponding data available in electronic form to interested users upon request.

The advent of data centres some 25 years ago – CDS was created in 1972 – made them appropriate places to deposit computer readable versions of long compilation catalogues. Eventually, the larger ones were only available in this electronic form, and were announced in the refereed journal with just a sample page (see, for example, the catalogue of Geneva photometric boxes by Nicolet 1981).

As computers were becoming an essential tool on astronomer's desks, the distinction between large compilation catalogues, and data tables published in astronomical papers faded out: individual tables in electronic form are now also commonly available from data centres; typically this concerns lists of observational results, reference data, calibrations to be used in models, etc.

Electronic usage of such catalogues and tables is made far easier when convenient format and documentation are organized.

The common format adopted by astronomers for complex datasets, images, and binary data is the FITS (for *Flexible Image Transport System*) format (Wells *et al.* 1981 and Grosbøl 1991). The idea is to encode both the definition of the data and the data itself in a machine independent way. This format has proven very useful for the exchange of image data, and is being supported by appropriately configured World-Wide Web browsers.

In fact most tables have a much simpler structure, and can be stored as ASCII tables. In order to cope with the increasingly large number of data files, a new standard for the format and documentation of simple ASCII tables was proposed recently by Ochsenbein (1994). The corresponding catalogue descriptions can be read easily by humans and by computers. This new standard is now also shared with other data centres (such as NASA/ADC) and publishers (CD-ROMs of the American Astronomical Society). It is also a key for automated data handling, exchange, and distribution; it supports, for instance, conversion from ASCII format to FITS, and is used to build the relational tables used by the database management system of VizieR (see below).

Since January 1993, following an agreement with the Editors of *Astronomy & Astrophysics*, tables from the main journal and from the Supplements Series are deposited at CDS in a systematic way. They are made available by CDS for electronic distribution, as soon as the corresponding article is released, as described by Ochsenbein and Lequeux (1995). Their documentation complies with the above mentioned standard.

Tables from *Astrophysical Journal*, *Astronomical Journal*, and *Publications of the Astronomical Society of the Pacific* which are published in the CD-ROMs of the American Astronomical Society (see the chapter by P. Boyce in this book) are also made available on-line at CDS, thanks to an agreement with the Editors.

2.2.3. *The on-line catalogue service*

All most frequently requested catalogues —as well as the data tables from individual articles described above— are stored at CDS, together with their documentation, as a fully electronic archive accessible on-line. This was made possible through the collaboration between all data centres, and especially NASA/ADC which had in the meantime installed its own magnetic tape collection into a near-line system, and had produced a subset of most frequently used catalogues on CD-ROM.

The *"Astronomer's Bazaar"* (Egret & Ochsenbein 1994) is a World-Wide Web service allowing:

- to query the list of catalogues, by keyword, or in browse mode,
- to display the corresponding documentation,
- and to retrieve the complete electronic files (eventually compressed), in ASCII or FITS format, from the anonymous `ftp` space of the CDS server.

More than 1500 catalogues and tables, for a total of several Gigabytes of data are available through this interface.

Furthermore, the new *VizieR* catalogue service allows, for a large subset of this collection of catalogues and data tables, to retrieve individual data records from selections on table columns, position, or astronomical object names.

This catalogue browser results from a collaboration with the *European Space Information System* (ESIS) project of European Space Agency, at ESRIN, Frascati. The service will also, in the future, support links to external archives of ground-based or space observatories. This implies preparatory discussions on formats and interfaces which are already under way with, e.g., the European Southern Observatory (ESO), the Canada–France–Hawaii Telescope, and the space agencies.

2.3. THE ALADIN INTERACTIVE SKY ATLAS

ALADIN is a new project currently under development (see e.g. Paillou *et al.* 1994, Bonnarel *et al.* 1997 and Bartlett *et al.* 1997), with the aim of creating an interactive atlas of the digitized sky with good quality astrometric and photometric calibration. This service will allow the user to visualize on his/her own workstation digitized images of any part of the sky, to overlay entries from astronomical catalogues or user data files, and to interactively access the related data and information from the SIMBAD database for all known objects in the field.

This new tool will be particularly useful for multi-spectral approaches such as searching for counterparts of sources detected at various wavelengths, and for a number of applications related to the database quality control and the careful identification of astronomical sources.

The ALADIN project has set up collaborations with the major groups providing digitizations of sky surveys (see e.g. McGillivray 1994). The database currently includes the first Digitized Sky Survey (DSS-I) produced by the Space Telescope Institute, covering the complete sky, as a set of slightly compressed images. Images of the second epoch survey (DSS-II) are gradually being included. In addition, high resolution images of crowded regions of the sky (Galactic Plane and Magellanic Clouds) have been provided by the MAMA facility at the *Centre d'Analyse des Images*, located at Observatoire de Paris, and by SuperCosmos of the Royal Observatory in Edinburgh.

Currently (February 1997), ALADIN is only available from a limited number of sites; it will eventually become a public interactive tool, available through the data networks. Users will be able to work on the details of astrometric and photometric plate calibrations in order to extract the full information from the digitized plates.

Aladin is conceived as a sophisticated image server, requesting a specific client program on the user's side; a shorter version, will provide digitized images of small regions of the sky around a given position, defined by sky coordinates, or by the name of an object at the centre, through a simple World-Wide Web server.

2.4. YELLOW-PAGE SERVICES

A growing number of astronomical resources are made available through the Internet. They sometimes contain much valuable information, frequently hidden in a deluge of non pertinent documents. Combining yellow-page services and meta-databases of active pointers may be an efficient solution to the data retrieval problem.

The CDS is hosting the following databases:

- **The StarPages** is the generic name for a collection of directories, dictionaries and databases which is described in more details in another chapter of this book by A. Heck who has been buiding up their contents for more than twenty years (see also Heck 1995). These very exhaustive datasets constitute a gold mine for professional, amateur astronomers, and more generally all those who are curious of space-related activities, and want to locate existing resources.
- **AstroWeb** (Jackson *et al.* 1994) is a collection of pointers to astronomically relevant information resources available on the Internet. The

browse mode of AstroWeb opens a window on the efforts currently developed —sometimes not very fruitfully, one should recognize— for making astronomically related, and hopefully pertinent, information available on-line through the World-Wide Web.

Yellow-page services are complemented at CDS by a specific effort for documentation and user support. An important aspect of this activity is the Reference Dictionary of Nomenclature, mentioned in Section 2.1.3.

2.5. BIBLIOGRAPHY AND LITERATURE SEARCH

Bibliography of astronomical objects is one of the unique features of the SIMBAD database. This service is currently being extended, in the context of new digital library services, in order to provide the user with a wider perspective of the current astronomical literature.

CDS has developed its own approach as a data centre contributing, from the origin, to bibliography services, and, more generally, to electronic publishing.

In the following we describe some of the steps taken in the recent years. Most of the features and resources mentioned below are made available not only as genuine CDS services, but also through more general digital library services, such as the *Astrophysics Data System* (ADS; see the chapter by Eichhorn in this book) —which is presently the major source for astronomical abstracts. It is, however, important to understand the specific contribution of data centres, which, although intrinsically with low visibility, brings a number of critical value-added features.

Bibliographic reference code   The bibliographic 19-digit reference code (Schmitz *et al* 1995) also shared with NED and ADS, is the common index (*bibcode*) for all references to published articles.

Simbad name resolver   As shown above, the cross-identification of a number of astronomical sources is a necessary step in order to cope with the variety of aliases and naming procedures. The Simbad name resolver is already used by ADS, and by major observatory archives.

Simbad bibliography   This creates the links between objects and related papers, a unique feature of SIMBAD. Reversely, SIMBAD can provide the full list of astronomical objects mentioned in a given paper, as well as links to basic data for these objects. This feature is also made available through ADS, or from electronic publication services (such as, e.g., *New Astronomy*).

On-line data tables The data tables are provided together with a standard documentation; the *bibcode* is the common key between papers and related data tables.

Abstracts of recent papers Through agreements with the Editors, the abstracts of *Astronomy & Astrophysics*, main Journal and Supplement Series, as well as the abstracts from the *Publications of the Astronomical Society of the Pacific*, are made available on-line a few weeks before publication. The collection of abstracts from these two journals, starting from January 1994, can be queried, on the World-Wide Web, by keywords or author names, and is also made available through ADS.

Mirror copy of ADS Since January 1997, the CDS has been hosting the European mirror copy of ADS, thanks to the support of the French Space Agency, CNES —together with the reciprocal installation of a SIMBAD clone in the USA, with NASA support, on the ADS site at the Center for Astrophysics.

Improving data centre work As a result from the collaboration with journal editors, CDS can now directly use the Table of Contents from the major journals to prepare the database updates.

It has also been shown that statistical analysis of bibliography linked to astronomical objects is an efficient tool for a large variety of scientific investigations and database quality control (Lesteven 1995).

3. Future perspectives for a global astronomical data and information service

The various activities of the CDS, as described above, are the pieces of a puzzle, all directed towards the organization of a simple and unified view of astronomical data and information. The developments of the World-Wide Web and the Hypertext Markup Language provide a sound basis for integrating complementary services, eventually distributed on more than one site, into a consistent system.

In order to achieve this goal, the CDS is gradually building the needed links between the different CDS services with the objective of providing a powerful information system to the users.

These new developments imply a dedicated effort in terms of research, in order to derive the best solutions according to the current information technology.

3.1. INTEGRATION OF DISTRIBUTED SERVICES

One main trend is the increase of interconnections between distributed on-line services, the "Weaving of the Astronomy Web" (following the title of a Conference organized in Strasbourg in April 1995 by D. Egret and A. Heck and published in a special issue of *Vistas in Astronomy*).

For the CDS, this means, in a first step, working towards integrated access to all CDS services.

3.1.1. *Generating Universal Links*

The need to build a transparent access to the whole set of CDS services has now become obvious, because of the user acquaintance with the easy navigation permitted by hypertext tools. ALADIN was the first development to give a comprehensive simultaneous access to both SIMBAD and the Catalogue service contents through a dedicated user interface.

To go further, the CDS is currently building a general data exchange model: GLU (Générateur de Liens Universels, Universal Link Generator; see Wenger *et al.* 1997) for managing links between heterogeneous services.

Such links cannot be hardcoded in the database, due to the need for evolutive addresses (URLs). A dedicated software uses a data dictionary to add linking information to the data which have been previously tagged. The GLU dynamically replaces tagged data by actual Internet locators (URLs), according to data type, data value, requested result, and even response time when mirror sites are involved.

This new tool will allow to take into account all types of information available at the Data Centre, or at any cooperating institution. The distributed nature of the dictionary will help linking external services together.

This approach is already used in the first Web version of SIMBAD (Wenger *et al.* 1997), and in the prototype bibliography service, where it gives access to a number of information pieces of scattered in the CDS services: SIMBAD references, *bibcodes*, and links to objects; tables from the Catalogue service; abstracts —with also links to the ADS service; Dictionary of Nomenclature; digitized images from ALADIN. The catalogue interface will soon be included in the same scheme.

In the medium-term range, the CDS World-Wide Web service will provide an integrated access to all CDS services, eventually, in some cases, in a "lite" version, i.e. with simpler functionalities than the dedicated interfaces (e.g. for ALADIN or SIMBAD).

More generally, with the development of the Internet, and of a large number of on-line services giving access to data or information, it is clear that universal tools giving access to distributed services are needed. This was, for instance, the concern expressed by NASA through the AstroBrowse workshop (Murray & Hanisch 1996). CDS, viewed as an astronomical *information hub*, and the other astronomical data centres, have their part to play in this context.

3.1.2. *Astronomical Server URLs*

On the technical side, the strong potential of the hypertext links, easy to install and ergonomic, is now well demonstrated, and the use of Hypertext Transfer Protocol will allow to implement transparent access to remote services.

One example, at CDS, is to complement *VizieR* with accesses to distributed archives and databases. The CDS plans are to include the observation logs of major observatories in the system: the user will thus be able to find out that the data he or she needs is available from a given dataset. The second step is then to provide active anchors towards the distributed archives, to allow retrieving the actual data.

A standard syntax for information exchange among remote astronomical services is currently discussed (see Ochsenbein *et al.* 1996) between CDS (Strasbourg, France), European Southern Observatory and Space Telescope European Coordinating Facility (Garching, Germany), Canadian Astronomy Data Centre (Victoria, Canada), Osservatorio Astronomico Trieste (Italy), and IUE Observatory (Vilspa, Spain).

The implementation of the links will be discussed on a case by case basis with the data providers.

3.2. TOWARDS NEW INNOVATIVE SERVICES

All the examples above have shown how data centres can be part of the electronic publishing environment. The next step will be to build up effective links between the journals and the databases, i.e. between the full research results and complementary information such as data, bibliographic references, images, and actual observations stored in observatory archives.

The ultimate goal is to provide the scientist with a full working environment integrating published results, data tables, calibration and data reduction software, images, etc. First practical examples are, for instance, to link the object name cited in a paper, with the SIMBAD information about the object; to link sky coordinates to the related area of digitized atlas, etc.

The door is now wide open to build up innovative links and services using this new opportunity. This will certainly have some consequences on the refereeing process, since it allows new possibilities for checking the paper contents, as well as on the internal data centre procedures.

New tools to manage textual information have to be developed, such as those shown by Lesteven *et al.* (1996). In this context, there is a very good opportunity for the Data Centre to give access to a new world of scientific information, to play its part in the definition of the links and of the new services, with an expert eye, in tight partnership with the journal editors and data providers.

Increased cooperation with all data providers will be critical in this new era, and the announcement of the *Universal Research Archive of Networked Information in Astronomy* (URANIA) is a clear demonstration that the astronomical community is now ready to take this step.

4. Conclusion

With the emergence of fully electronic publication, new digital library services are being organized, and pave the way to innovative new services, linking publications to information from other sources, and making use of new methods for textual information retrieval.

The data centres expect to play a key rôle in these new developments, taking advantage of their expertise in the development of value-added services, and of their long term involvement towards a fully linked astronomy information system. In the same time, their commitment for the long-term archiving of data makes them also natural places for answering the needs of a long-term archive of all electronic publications.

References

1. Bonnarel, F., Ziaeepour, H., Bartlett, J.G., Bienaymé, O., Créze, M., Egret, D., Florsch, J., Genova, F., Ochsenbein, F., Raclot, V., Louys, M., Paillou, Ph. (1997), "The Aladin Interactive Sky Atlas", in IAU Symp. 179, New Horizons from Multi-Wavelength Sky Surveys, *in press*
2. Bartlett, J., Bonnarel, F., Egret, D., Genova F., Ziaeepour, H., Bienaymé, O., Ochsenbein, F., Crézé, M., Florsch, J., Louys, M., Paillou, Ph. (1997), "The Aladin Project: A Tool for Multiwavelength Cross-identifications", Data Analysis Workshop, Erice, October 1996, *in press*
3. Egret, D., Wenger, M., Dubois, P. (1991), in *Databases & On-line Data in Astronomy*, Albrecht & Egret (Eds.), Kluwer Acad. Publ., pp. 79–88.
4. Egret, D., Ochsenbein, F. (1994), CDS Inform. Bull. **44**, 57.
5. Egret, D., Crézé, M. , Bonnarel, F., Dubois, P., Genova, F., Jasniewicz, G., Heck, A., Lesteven, S., Ochsenbein, F., and Wenger, M. (1995a), "A global perspective on astronomical data and information: the Strasbourg astronomical data centre (CDS)", in *Information & On-line Data in Astronomy*, Egret & Albrecht (Eds.), Kluwer Acad. Publ., pp. 163–174.
6. Egret, D., Genova, F., Dubois, P., Heck, A., Lesteven, S., Ochsenbein, F., Crézé, M., Bonnarel, F., Jasniewicz, G., and Wenger, M. (1995b), Vistas in Astronomy **39**, pp. 195–202.
7. Egret, D., Heck, A. (1995), Vistas in Astronomy **39**, pp. 1-121.

8. Genova, F., Bartlett, J. G., Bienaymé, O., Bonnarel, F., Dubois, P., Egret, D.,
    Fernique, P., Jasniewicz, G., Lesteven, S., Monier, R., Ochsenbein, F., and Wenger,
    M. (1996), "CDS as an Astronomical Information Hub", Vistas in Astronomy **40**,
    pp. 429–437.
9. Grosbøl, P. (1991), in *Databases and On-line Data in Astronomy*, M.A. Albrecht &
    D. Egret (Eds.), Kluwer Acad. Publ., pp. 253–257.
10. Heck, A. (1995), "The Star*s Family – An example of comprehensive yellow-page
    services", in *Information & On-line Data in Astronomy*, D. Egret and M. A. Al-
    brecht, Eds., pp. 195–205.
11. Jackson, R., Wells, D., Adorf, H.M., Egret, D., Heck, A., Koekemoer, A., Murtagh,
    F. (1994), "AstroWeb - A Database of links to astronomy resources", *Astron. As-
    trophys. Suppl.* **108**, pp. 235–236.
12. Laloë, S., Beyneix, A., Borde, S., Chagnard-Carpuat, C., Dubois, P., Dulou, M.R.,
    Ochsenbein, F., Ralite, N., Wagner, M.J. (1993), *CDS Inform. Bull.* **43**, 57.
13. Laloë, S. (1995), *Vistas in Astronomy* **39**, pp. 259–270.
14. Lesteven, S. (1995), "Multivariate data analysis applied to bibliographical informa-
    tion retrieval: SIMBAD quality control", Vistas in Astronomy, **39**, pp. 187–193.
15. Lesteven, S., Poinçot, Ph., Murtagh, F. (1996), Vistas in Astronomy, **40**, pp. 395–
    400.
16. Lortet, M.C., Borde, S., and Ochsenbein, F. (1994), *Astron. Astrophys. Suppl.* **107**,
    pp. 193–218.
17. MacGillivray, H.T. et al., Editors (1994), "Astronomy from Wide-Field Imaging",
    Postdam, Germany, Kluwer Acad. Publ., pp. 1–760.
18. Murray, S.S. and Hanisch, R.J. (1996), *AstroBrowse* Workshop Report, communi-
    cation at ADASS '95, unpublished
19. Nicolet, B. (1981), Astron. Astrophys. Suppl. Ser. **48**, pp. 485–490.
20. Ochsenbein, F. (1994), *CDS Inform. Bull.* **44**, pp. 19-28.
21. Ochsenbein F., Lequeux J. (1995), Vistas in Astronomy, **39**, pp. 227–234.
22. Ochsenbein, F. et al. (1996), *Astronomical Server URL*,
    http://vizier.u-strasbg.fr/doc/asu.html
23. Paillou, Ph., Bonnarel, F., Ochsenbein, F., Crézé, M. (1994), in *Astronomy from
    Wide-Field Imaging*, Postdam, Germany, H.T. MacGillivray Ed., Kluwer Acad.
    Publ., pp. 347–351.
24. Schmitz, M., Helou, G., Dubois, P., LaGue, C., Madore, B., Corwin, H.G. Jr, and
    Lesteven, S. (1995) "NED and SIMBAD conventions for bibliographic reference
    coding", in *Information & On-line Data in Astronomy*, Egret & Albrecht (Eds.),
    Kluwer Acad. Publ., pp. 259–270.
25. Wenger, M., Fernique, P., Genova, F., Bartlett, J.G., Bienaymé, O., Bonnarel, F.,
    Dubois, P., Egret, D., Jasniewicz, G., Lesteven, S., Monier, R., Ochsenbein, F.
    (1997), BAAS **189**, #06.02.
26. Wells, D. C., Greisen, E. W., Harten, R. H. (1981), *Astron. Astrophys. Suppl.* **44**,
    363–370.

THE DIGITAL LIBRARY OF THE ASTROPHYSICS DATA SYSTEM

G. EICHHORN
Smithsonian Astrophysical Observatory,
60 Garden Street, MS-83
Cambridge MA 02138, USA
gei@cfa.harvard.edu
http://hea-www.harvard.edu/~gei/geichhorn.html

1. Introduction

The Astrophysics Data System (ADS) provides access to astronomical bibliographic information, including references, abstracts, and full journal articles, as well as links to other on-line information sources like on-line electronic journals and on-line data.

This section will first provide a brief history of the ADS, a general introduction of the current system, and a more detailed description of some of the parts of the ADS.

2. History

The ADS started as a system to access the various data repositories containing data from NASA's space missions. In the late 1980s NASA asked the astronomical community for an evaluation of the status of data access (Squibb 1987). The recommendation was that a system be established that allows access to the data at the various data centers through one system (Squibb 1988). In 1991 an operational version of such a data access system was released (Good 1991, Murray 1993 and Eichhorn 1994). It was based on proprietary technology since at that time there was no public domain distributed data system available. This system allowed access to data at various collaborating data centers at the record level. It allowed users to query different databases and to retrieve data for display and further analysis.

Astrophysics and Space Science **247**: 189–210, 1997.
© 1997 *Kluwer Academic Publishers.*

In 1993 the abstract service was added to the ADS. It contained at the beginning about 160,000 abstracts. It allowed fielded queries with a sophisticated relevance scoring system (Kurtz 1992 and Eichhorn 1995a).

In 1994 the emergence of the World Wide Web (WWW) made this custom built distributed data system obsolete. The open architecture of the WWW allowed the data centers to bring their data on-line easily and to provide search capabilities tailored to their particular needs.

The ADS abstract service was the first part of the ADS that was interfaced to the WWW in February 1994. A WWW catalog interface tool followed quickly (Eichhorn 199b and Accomazzi 1995). With the introduction of the WWW access tools to the abstract service, the usage tripled within one month. Since then the usage has been increasing almost continuously. Fig. 1 shows the number of users per month since the start of the abstract service in 1993. It shows the fairly slow increase during the first year with the classic ADS interface and the rapid growth after the introduction of the WWW interface.

Distinct Users, Abstract Service

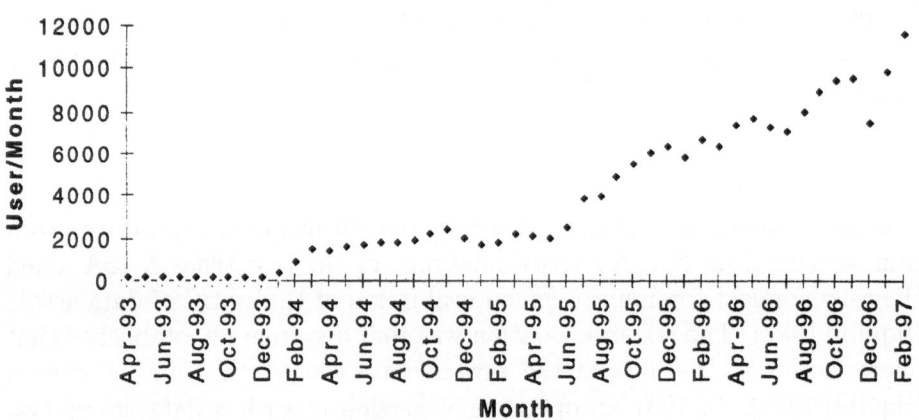

Figure 1. Number of users of the ADS abstract service as a function of time. The increase of users in February 1994 is due to the introduction of the WWW interface.

With the WWW technology other features became feasible, namely linking between different data systems. Very quickly the ADS made use of this technology and started linking data at different data centers to the references in which these data are described.

The first such links were from the abstract service to the SIMBAD database at the *Centre de Données astronomiques de Strasbourg* (CDS) (Egret

1991). Queries for astronomical objects would be sent to the SIMBAD server at the CDS. The list of references that was retrieved would then be matched against the ADS database and the matching references returned to the user. Subsequently links to the data tables at the CDS, the General Catalogue of Photometric Data, maintained at the University of Lausanne, Switzerland (Mermilliod 1996), and the Astronomy Digital Image Library (ADIL) project at the National Center for Supercomputing Applications (NCSA) (Plante 1996), were included where appropriate in the returned ADS references. Since then the SIMBAD project has provided the necessary information to us to include links to lists of objects that are mentioned in journal articles. This now allows our users to directly get information about objects mentioned in retrieved journal references.

With a large part of the astronomical literature references and abstracts available through the ADS, we proceeded to make full journal articles available on-line. After obtaining permission from the copyright holders of the various journals, we scanned these journals. Scanning was done at 600 dpi to ensure highest quality display and printing capabilities. This service came on-line in January of 1995 (Accomazzi 1996). It currently provides access to 11 journals for various time spans. The near term goal is to have complete coverage for these journals from 1975 to present. The longer term goal is to provide complete coverage for all these journals (and hopefully more) from Volume 1 to present.

In order to protect the revenue of the journal, we make the latest issues available only after a grace period whose length depends on the preference of the journal. This can be as short a 6 months to as long as 2 years.

In 1996 the Astrophysical Journal Letters (ApJL) came on-line in electronic form (Boyce 1996). More journals will be on-line in 1997. The ADS provides links to these on-line articles. The on-line journals in turn utilize the ADS abstract service to make links to abstracts available directly from their reference lists in the on-line articles. This close cooperation between the different data and information systems in Astronomy is the first example of a complete digital library that allows the user to easily move from one information source to another and to quickly find related data. Fig. 2 shows schematically the Astronomical Web of information sources. The collaboration of different groups is now called Urania (Universal Research Archive of Networked Information in Astronomy).

3. Current ADS System

The ADS has two major parts, the abstract service and the on-line journal articles. This section will give an overview over the two parts and will describe some parts of the system in more detail.

Number of Abstract Queries

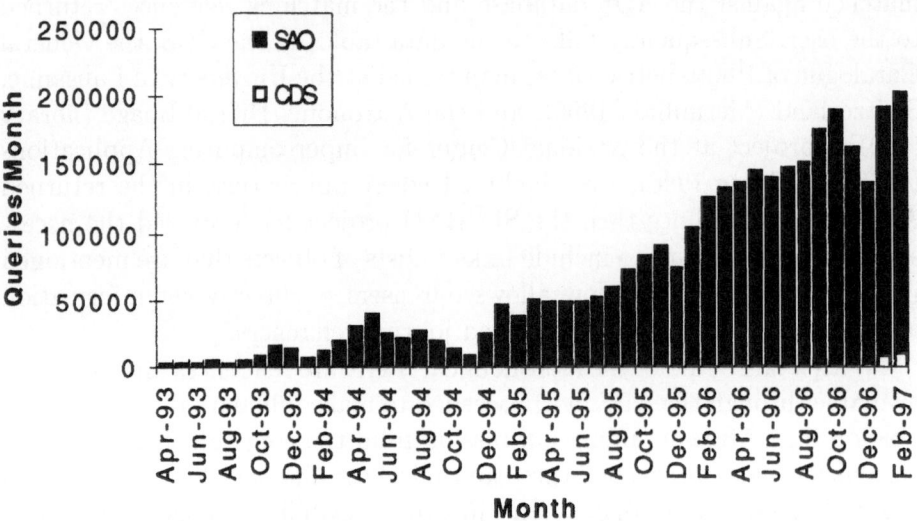

Figure 2. Interlinking of the various parts of the Digital Library in Astronomy. This cooperation is called Urania (Universal Research Archive of Networked Information in Astronomy).

3.1. ADS ABSTRACT SERVICE

3.1.1. *General Description*

The abstract service provides access to abstracts and references of most of the astronomical literature since 1975. The references are accessible in three databases: Astronomy, Space Instrumentation/Optics, and Physics/Geophysics. As of January 1997, the astronomy database contains about 250,000 references, the Instrumentation database has almost 500,000 references, and the physics database about 250,000 references. Keeping separate databases for different topics improves the quality of the information returned from the search engine.

The databases are currently updated approximately every two weeks. In the near future a quick update capability will be implemented that will allow much more frequent partial updates (one or more per day). Full reindexing of the complete database will still be required every few weeks. This new capability is expected to be available in March 1997.

Most of our data came from the NASA Scientific and Technical Information service. This project was abstracting all the astronomical literature, including journals, conference proceedings, PhD theses, NASA internal reports, etc. In 1995, NASA/STI discontinued abstracting most of the literature. Since then we are receiving abstracts or at least references (titles and author lists) from most journals directly. For the Astronomy database we receive abstracts from all major journals directly. For the Instrumentation database we receive all conference proceedings abstracts that the Society for Optical Engineering (SPIE) publishes.

The databases can be searched with a sophisticated search engine, allowing fielded searches. The retrieved references are ranked according to their relevance to the query, taking into account word frequencies for weighting.

The database is indexed such that retrieval of abstracts that contain synonyms for the search words (e.g. M31 and Andromeda) is possible. The list of synonyms was compiled manually taking into consideration expert knowledge of the technical language used in Astronomy. This synonym list has proven to be one of the most valuable assets of the ADS because it greatly improves the quality of the searches. For added flexibility, the user has the option of turning off the automatic synonym replacement.

Parsing of the abstracts before indexing and of the user input before searching allows further refinements of the search capability. For instance the words "Be Star" are indexed together when they appear in this sequence, making it possible to find references about Be Stars. Searches for the two words Be and Star separately would not give useful results.

3.1.2. Search Engine

General.   The search engine for the abstract service was written specifically for this application. This allowed us to implement optimizations specifically tailored to this type of data. It allows us to quickly implement suggestions from our users to improve the quality, ease of use, and speed of the searches. Separate indexes for authors, title words and text words are created during indexing. Additionally, indexes for word pairs are built to enable searching for multi-word phrases. All searchable tables (index tables for each field, list of bibliographic codes, etc) are maintained in memory. When a new search is initiated. the software attaches to the tables in memory. No file accesses or loading of tables has to be done. This provides significant execution speed improvements compared with conventional databases. All tables in memory are sorted. For correlations that need to be searched in both directions. inverted index tables are held in memory also. This means that all searches can be done as binary searches.

During a search, the database is searched for the specified words in each search field. The results are then combined according to the user specified logic. For each field, the following logic options are available:

- Combine with 'OR': A reference is returned when at least one of the words specified in the field was found.

- Combine with 'AND': A reference is returned only when all words specified in the field were found.
- Simple Logic: This allows the user to specify for each word individually whether it is optional for return, required, or must not be in the returned reference. Required words have to be prefixed with a '+', words that must not be present are prefixed with '-'. Words without a prefix are optional.
- Complex Logic: This mode allows the user to specify full boolean expressions containing 'and', 'or', 'not', and parentheses for grouping.

For all logic options, the following constructs are recognized:

- Prefix '=': This specifies that no synonym replacement should be done for this word. Normally a search is done for a word and all its synonyms.
- Prefix '#': If synonym replacement has been turned of, this prefix specifies that synonym replacement should be done for this particular word.
- Phrase Searching: The search engine can search for sequences of words. This allows for instance to search for "black hole" as a phrase, rather than searching for 'black' and 'hole' separately. Phrases can be specified by enclosing them in single or double quotes, or by concatenating the words with '.' or '-' (without any blanks in between).

Once the searches for each field are completed, and the results within each field are combined as per the logic selected, the results from the different fields are combined. Again, the user can determine how this is done. Normally, references are returned if at least one field shows a match. The user can select one or more of the fields to be required for the result. If this option is selected for a particular field, only references that have at least some match for this field will be returned. This allows for instance to find references by a particular author by requiring the author field to be present. The other fields would then be used to further select and sort references from this set.

Scoring of the retrieved references. Each selected reference is assigned a score that shows how closely this reference matches the query. By default, the score is calculated for each field from the number of items matched. For the author field and the object field, the score is the number of matched items divided by the total number of items in the search field. For the title and text fields the score is determined as the sum of the weights for each matched word divided by the sum of the weights for all words specified in the field. The weight of a word is $1/\log 10(\mathrm{freq})$ where freq is the frequency of the word in the database. This scoring algorithm weighs less frequent words more, since they are presumably better indicators of the relevance of an abstract to a particular subject. This frequency weighted scoring for title and text fields can be turned off if desired.

The total score is finally determined by multiplying the score for each field with the weight for this field and adding up these products. The weights for each field can be adjusted by the user. This allows the user to for instance put more weight on the object field if this seems important. By specifying a weight of -1.0 in a particular field, this weighting scheme can be used to select against a particular field. If for instance an author is specified, together with words in the text field, and the weight of the author field is set to -1.0, only references that have matches in the text field, but do NOT contain this author in their author list are returned.

The final score is then normalized by the total possible score for the query. A final score of 1 means that all items in the query were matched by this reference.

Filters. The final output can be further modified with several filters. These filters let the user select further attributes of the returned references. The following filters are available:

- Publication date: Selects references with a publication date in the specified range.
- Entry date: Selects references that were entered into the database after a specified date.
- Minimum score: Selects references whose score is larger than the minimum score specified.
- Bibliographic sources: Selects references from specified sources (refereed journals only, non-refereed journals only, specified journals only).
- Available data: Selects references with specified data links (e.g. full articles, or electronic on-line version, etc).
- Groups: Selects references from the bibliography of the specified group.

3.1.3. *Returned Information*

The result of a search is a list of references. The list contains the bibliographic code, the authors, the title, the publication date, the score and a list of links.

Bibliographic Codes. The bibliographic codes are a very important part of the whole digital library in Astronomy. They allow the different systems to work together by uniquely identifying every reference. They are described in detail by Schmitz *et al.* (1995).

These codes are 19 characters long. They have the following form:

where YYYY is the year, JJJJJ is the abbreviation for the journal (e.g. ApJ, AJ, MNRAS, Sci, PASP, etc.), VVVV is the volume number, M is used when needed to indicate "special" issues (such as "L" for Letters, "P" for pink pages), PPPP is the page number, and A is the first letter of the first author's surname. The fields are padded with periods (.) so that the code is always 19 characters long. The journal is left-justified within its 5 characters, and the volume and page are right-justified. A list of journal abbreviations already in use is available at:

http://adsabs.harvard.edu/abs_doc/journal_abbr.html.

These bibliographic codes can be generated automatically from a regular reference in the reference section of a journal. This allows the electronic journals for instance to generate links to abstracts in the ADS in their reference section. We provide a bibcode verification tool that allows the journal editors to check whether the bibcodes that they generated are indeed valid bibcodes.

These bibcodes are the glue that holds the digital library together and allows different parts of the system to easily interact with other parts.

Links to other information.   One of the most important aspects of a distributed digital library on the Web is the cross-linking between different parts of the library. Therefore the generation and maintenance of links from the ADS references to other information connected with each reference has very high priority. These links can be accessed directly from the retrieved references. They are anchored to a string of letters. Table 1 gives links, anchored to the specified letters, that are currently available.

Following is a more detailed description for some of these links.

Abstracts ('A' and 'O' links).

These links provide access to the full information available in the ADS. This includes the title, author list, journal information, publication date, author affiliation, keywords, and the full abstract (where available). The returned abstract page also contains links to all the information mentioned in Table 1, as well as other relevant links like links on author names to a list of email addresses, current affiliations and the StarPages (Heck, 1997), and a mechanism to use the current abstract as input to a new query to find related articles.

References and Citations ('R' and 'C' links).

In late 1996, the American Astronomical Society (AAS) purchased a list of references from the Institute for Scientific Information (ISI). These new data allowed us to build the list of references and citations for most journal

TABLE 1. Available links.

Link Letter	Information available at this link
A	Abstract provided by NASA/STI.
O	Original author abstract provided by the journal or author.
F	Full article available through the ADS article service as scanned images.
E	On-line electronic version of the journal article, usually at the site of the publisher. Access to the electronic version may be restricted to subscribers of the journal.
P	PDF version of the article. This too may be restricted to subscribers of the journal.
D	On-line data from the reference in electronic format.
S	List of astronomical objects referred to in the article (provided by the SIMBAD database).
M	The article can be mail-ordered from a document delivery service. This service is fee based and is handled directly through the document delivery service.
I	Additional information about the article provided by the author.
R	List of references for this article. It generally includes only the references to other journal articles.
C	List of references that cite this article.
T	Table of contents for this reference (for books and proceedings volumes).

articles included in the Astronomy database. It includes only references to journal articles, not to conference proceedings. Each reference for which we have this information has a link to the set of references in that article and to the set of references that cite this article.

On-line Electronic HTML and PDF articles ('E' and 'P' links).

These links allow our users to view on-line journals in HTML format and to download PDF files for printing. We are currently working closely with the AAS and University of Chicago Press, which publish the Astrophysical Journal, to provide direct links to their on-line articles. The AAS has agreed to use the same bibliographic codes (see above and Schmitz 1995) that we use to identify articles. This makes it very easy to provide these links automatically right after publication. Other on-line journals are using the same codes or have indicated that they will use them when they come on-line.

On-line Data ('D' links).

One important aspect of the on-line digital library is to provide access not only to the literature, but also to on-line data. Traditionally it has been very difficult to get data from an article for further analysis. Hand typing the data was the only choice in most cases, which is error prone. The ADS is working with several data centers to provide links from the references to the data associated with these articles. This allows our users to directly access these data tables and download them in electronic format for further analysis.

Document Delivery Services ('M' links).

These links allow our users to order specific articles from a document delivery service. All such transactions are handled directly between the user and the delivery service. Our M-link provides the information about the article to the delivery service. We currently have such an arrangement with SPIE. All their articles can be ordered through their delivery service. Our link to the SPIE delivery service contains the information for the article that is being ordered. They then return the order form for that particular article and ask the user to supply shipping and payment information. We are discussing similar arrangements with other document delivery services.

Table of Contents. For a considerable number of journals we receive the table of contents (ToC) on a regular basis. For these journals we provide direct access to the ToCs. One of our pages provides access to either the latest published volume of each journal or the last volume that the reader has not yet read, depending on the settings of the user preferences. The last read volume for each of the major journals is remembered through the WWW cookie preference system (see next section about "Customizing the ADS"). This works only for users whose browsers handle WWW cookies. For the other users, this page always returns the latest published volume. There are two versions of this page available, one with images of the journals for easy journal selection, and one with plain text links for faster downloading. The ToCs for older volumes can be retrieved through a second ToC query page. That page allows the user to retrieve ToCs for selected journals by either volume or publication year and month. For older volumes there may not be all articles for a given volume in our system.

3.1.4. *Customizing the ADS*

Most frequently used WWW browsers now support a scheme called "Cookies". This scheme allows a server to uniquely, but anonymously, identify a user. When our server communicates with a cookie-aware browser, we set a so-called cookie. This is just a unique string that will later identify his user.

The browser stores this cookie on the users local machine together with the information about which host has sent that cookie. In all subsequent transactions, the browser will include this cookie with any information that is sent to this host. This allows the ADS server to determine from which user such a request originated. We provide a form that allows our users that have cookie-handling browsers to set certain preferences in searching and returning information. These preferences are stored in a database on the ADS server. They are used to customize what information is returned and how it is returned. The preference settings form allows the user for instance to select the default resolution for printing articles and how to return them (to printer directly, to file, via fax or email). It allows the user to customize font sizes and colors for the returned pages. The user can also store a default query form that can be tailored to the needs of the particular science subjects. We intend to continue to exploit this capability to make the use of the ADS more convenient for our users.

3.1.5. *Alternate Access Methods*

Usually, the ADS databases are accessed through WWW forms. However, there are other possibilities for using these databases:

Direct Hyperlinks.   Direct links to individual abstracts and articles can be embedded in HTML documents. These links have the form:
`http://adsabs.harvard.edu/cgi-bin/bib_query?bibcode`
where bibcode is the bibliographic code for the requested reference. Such a link will return the formatted abstract with links to all available information.

Direct links to journal articles have the form:
`http://adsabs.harvard.edu/cgi-bin/article_query?bibcode.`

Embedded Queries.   Such embedded query hyperlinks in documents perform ADS abstract service queries. This allows users for instance to include links in documents that return their bibliography. Such links have the form:
`http://adsabs.harvard.edu/cgi-bin/abs_connect?parameters=values`
`        &return_req=no_params.`
The parameters describe what query is to be executed. A description of these parameters is available at:
`http://adsdoc.harvard.edu/abs_doc/embed_form.html.`
An example of such a link is:
`http://adsabs.harvard.edu/cgi-bin/abs_connect?author=eichhorn,+g`
`        &return_req=no_params.`
This link returns the bibliography for the author of this section.

Perl Scripts. We provide Perl scripts that allow developers to directly access our database and retrieve information in Perl arrays. This allows other systems to build their own query forms and custom format the returned abstracts. The Perl scripts and a description on how to use them are available at:
`http://adsdoc.harvard.edu/abs_doc/direct_access.html`.

Email Queries. The ADS can be queried through email without needing WWW access. The results are returned via email. This allows our users for instance to automatically query the ADS every month to get the latest entries in our database. The user can build a query form that produces the desired results and the save the form locally. Sending this saved query form via email to `adsquery@cfa.harvard.edu` will execute this query and return the results to the sender. By specifying -31 in the Entry Date Day field, the query will return matching references that were entered in the database within the last 31 days. A more detailed description of this capability is available at:
`http://adsdoc.harvard.edu/abs_doc/whatsnew_help.html`.

3.1.6. *Usage Statistics*

The ADS abstract service has seen rapid growth since the introduction of the WWW interface. In February 1997, the ADS was accessed from more than 10,000 hosts. This translates into many more users since many hosts (especially hosts of Internet access providers) are use by more than one user. These users made over 200,000 queries and retrieved 3.9 million references. Fig. 3 shows the evolution of the number of queries with time and Fig. 4 shows the number of references retrieved per month since the introduction of the abstract service in 1993. It clearly shows the increase in access after the introduction of the WWW interface in February 1994.

The number of people using the abstracts service (more than 10,000) is comparable to the number of astronomers world wide. This indicates that almost all astronomers by now use the ADS at one time or another. Out of the 10,000 users in February, about 2,000 users issued as many queries as there were working days in January (on average one query per working day). This means that about 20% of astronomers use the ADS on a daily basis.

Fig. 5 shows the types of queries that our users make. In January 1997, 65% of all queries used the author field, 29% the text field, 19% the title fields, and 6% the object field. 16% of the queries use multiple fields. These values are quite stable in different months.

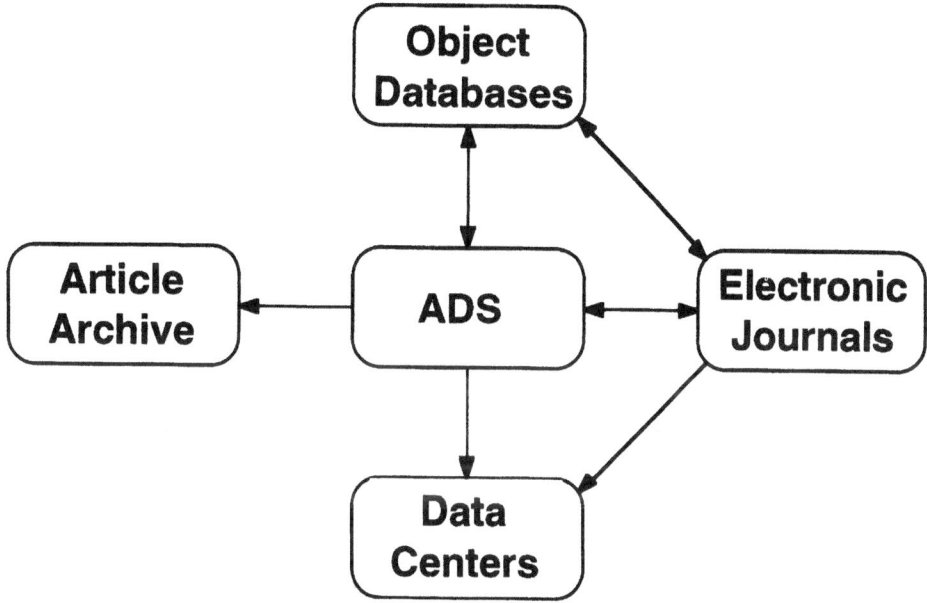

Figure 3. Number of queries as a function of time. The ADS mirror at CDS came on line in December 1996.

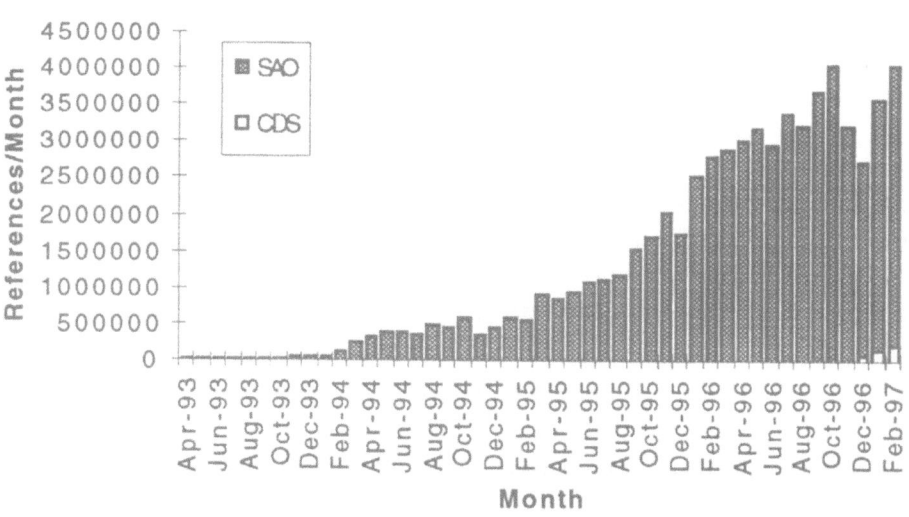

Figure 4. Number of references retrieved as a function of time.

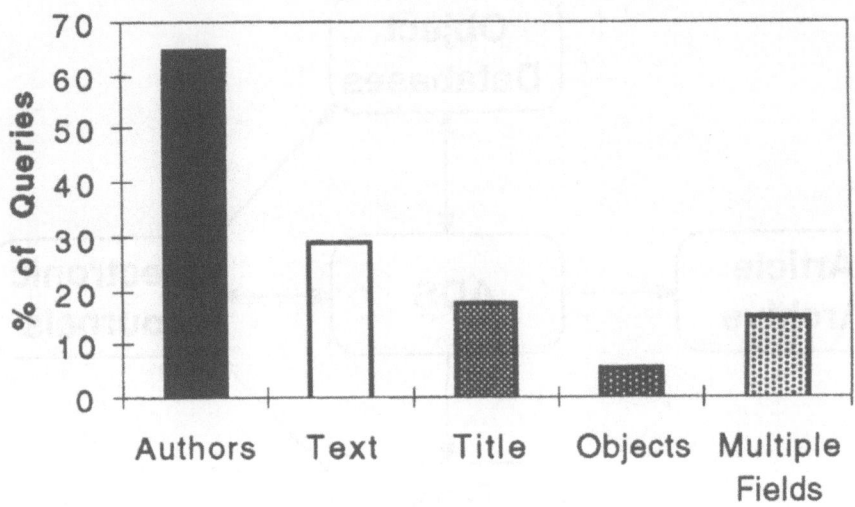

Figure 5.   Histogram of the types of queries used. The clear majority of queries are author queries (65%), followed by text queries. About 15% of the queries use more than one field. Between 5% and 6% of the queries use object names.

Clearly, author queries are by far the most important queries, followed by text queries. Fig. 6 shows how many queries use different numbers of words in each field. For author queries, the vast majority of queries use only one author. Only 15% of author queries use two authors, and only 1% use 3 or more authors. For object queries, 93% use only one object name. On the other hand, about 1/3 of the title queries use one word, 1/3 use 2 words, and 1/3 use 3 or more. In the text field, the largest number of queries uses two words (31%), 26% use one word, and the rest 3 or more words. There is a significant fraction of text queries that use 40 to 120 words. These are due to the feedback queries, where the abstract is used as a new query to find an exhaustive reference list about a particular subject. This capability is provided directly from the returned abstract and is therefore used quite frequently.

The ADS is used all day long. There was no 20 min period in January 1997 without at least one query, except for 3 periods just after midnight on January 1st on the east coast of the US and 2 periods when our server was down. Fig. 7 shows the usage for each hour of the day (Universal Time). You can distinguish the times when certain user groups start and stop using the ADS. But even at the lowest usage level, there are on average more than one query per second.

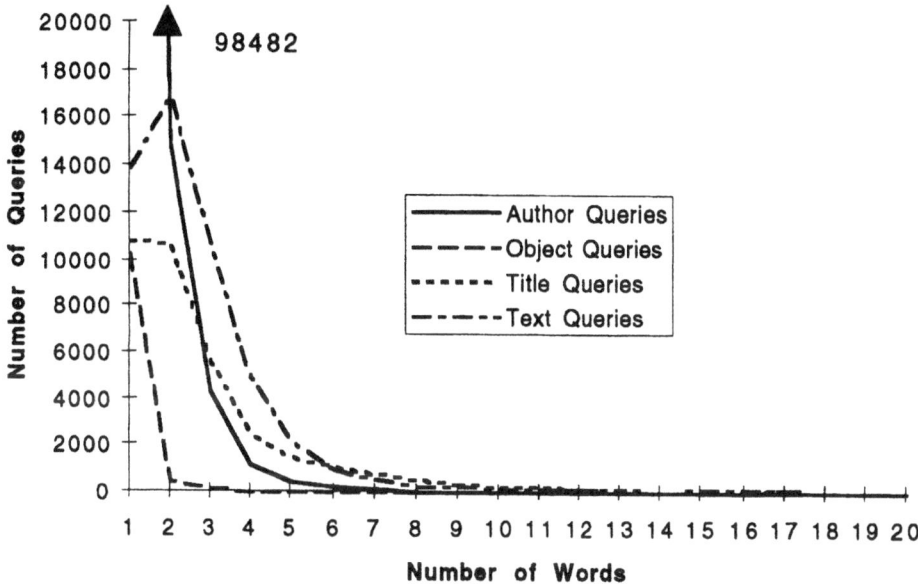

Figure 6. Histogram of the number of queries that use various numbers of words in the different query fields. For author queries and object queries, the vast majority of queries use only one word. For text queries, the majority uses two words in the text query field.

3.2. ADS ARTICLE SERVICE

3.2.1. *General Description*

The ADS article service provides access to scanned images of full journal articles. We have agreements with the publications listed in Table 2 to scan their journals and make them available on-line a certain time after publication:

Our scanning is not yet complete as of February 1997. We have scanned 700,000 pages so far and have about 400,000 pages on-line. About 200,000 pages are waiting to be scanned. We are also discussing similar arrangements with other journals. Currently the article database uses about 140 GBytes of disk storage. Eventually we will need up to 500 GBytes. We have decided to use regular magnetic disks for the storage of the article images. This is quite feasible with todays disk technology and will be even easier with the next generation of magnetic disks which should hold more than 20 GBytes per disk.

3.2.2. *Image Storage Technology*

The journals were scanned with a resolution of 600 dots per inch (dpi). This provides very high quality images. Printed with a 600 dpi printer,

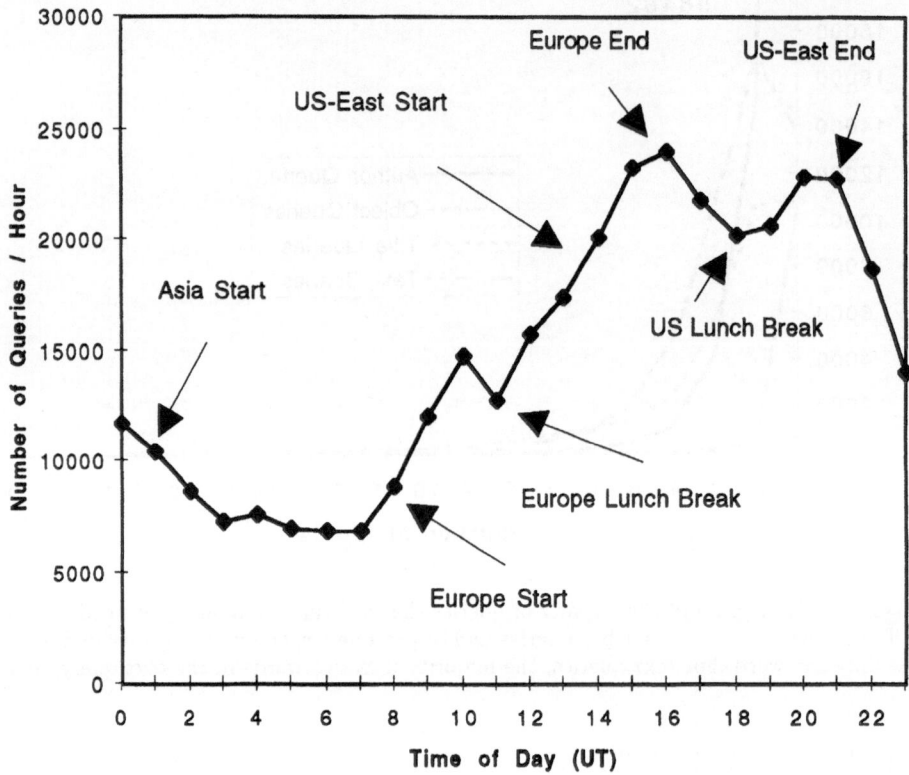

Figure 7. Daily usage pattern of the abstract service. It shows start and end of the workday for the different areas of the world.

TABLE 2. Publications available on line.

Astronomical Journal (USA)
Astrophysical Journal (USA)
Astronomy & Astrophysics (Germany)
Baltic Astronomy (Lithuania)
Bulletin of the Astronomical Society of India (India)
Contributions of the Astronomical Observatory Skalnate Pleso (Slovak Republic)
Monthly Notices of the Royal Astronomical Society (Great Britain)
Publications of the Astronomical Society of Australia (Australia)
Publications of the Astronomical Society of Japan (Japan)
Publications of the Astronomical Society of the Pacific (USA)
Revista Mexicana de Astronomia y Astrofisica (Mexico)

they are almost indistinguishable from the original. The images are stored on our server in two resolutions (200 dpi and 600 dpi). This allows the fast retrieval of a smaller, lower resolution version. The images are stored in TIFF format with G4 compression. This format provides a very good compression ratio. An added advantage is the fact that Postscript Level 2 uses the same compression algorithm. The .tiff files can therefore be directly converted into Postscript Level 2 files for printing.

The user can retrieve images in several formats. All these formats are generated in real-time when the user requests them:

GIF format, 100 dpi, 4 level gray-scale, anti-aliased for screen viewing. These images are about 100-200 KBytes per page in size. They download reasonably quickly. The quality and resolution is good enough so they can be read on the screen. To save disk space, but also conserve compute resources, the .gif images are created on demand and then cached. This allows for fast downloading of frequently requested articles, while saving space by not storing screen-view images of older articles that are read only infrequently. The time to create the .gif images in real-time is small compared to the transfer time and does therefore not significantly increase the time to download a page. The caching is done more to decrease the load on our server than to speed up the image transfer.

Postscript files for printing.   We provide 3 Postscript versions: 600 dpi Postscript Level 2, 200 dpi Postscript Level 2, and 200 dpi Postscript Level 1. All 3 versions are produced from the stored .tiff files in real-time. The 600 dpi version takes longer to download and to print, but it provides the best print quality. We store the 200 dpi .tiff files in order to make the retrieval of the low resolution version as fast as possible. This version prints very quickly. The quality is still better than a fax and generally as good as a good quality photocopy. The 200 dpi version can also be retrieved in Postscript Level 1 format. This is for back compatibility with older Postscript printers. This version is rather large, and takes long to download and print.

PCL files for printing on PC printers.   These versions too are created in real-time. They print on most PC printers. They are available at 300 dpi and 150 dpi (PCL does not allow for 600 dpi printing).

TIFF G4 compressed files.   The original .tiff files can be downloaded for viewing, printing or other processing.

One important note to our users: All these journal article images are copyrighted by the owner of the original copyright of the journal. They are to be downloaded only for personal use. Any commercial use is prohibited, unless express written permission of the owner of the copyright is given. Images should in general not be stored on the user's system and should not be distributed further. All access should be done only through the ADS article service.

3.2.3. *Accessing the Article Service*

The article database is closely linked with the ADS abstract service. Any retrieved abstract for which a scanned article is available is directly linked to the on-line article. The articles can also be retrieved directly by volume and page number. The user can retrieve articles either for viewing on-screen, for printing, or for storing on local disk. The ADS server will instruct the browser to handle the file according to the requested retrieval method. Most browsers can handle direct printing requests, so the file does not have to be stored and then printed. The user can select whether to print the complete article or only selected pages.

For users whose browser can handle WWW cookies, the system returns a form with just one default print button and a link to another page that has all the retrieval options. The default behavior can be set in the user preferences. This includes for instance the resolution, whether to return Postscript or PCL, and whether to send the file directly to the printer, to local storage, through fax or through email. For users whose browser does not handle cookies, a page will be returned that allows the selection of the most important of these options directly and all the others through another page.

3.2.4. *Statistics*

Fig. 8 shows the usage pattern of the article service. In 1995 we had only scanned and made available on-line the Astrophysical Journal Letters. By the end of 1995, more journals were added on a regular basis. In February 1997, over 235,000 pages were retrieved, with the usage continuously climbing, except for a December lull each year. The number of pages retrieved is more than what any astronomical library handles.

Fig. 9 shows the usage pattern as a function of age of the viewed article. The usage declines as $1/t$ with age. Clearly the most read articles are the most recent articles. However, the decrease with age seems to be slower than the decrease in usage of the on-line electronic Astrophysical Journal. This would indicate that the most recent articles are read mainly on-line directly at the journal and found from the table of contents, whereas the

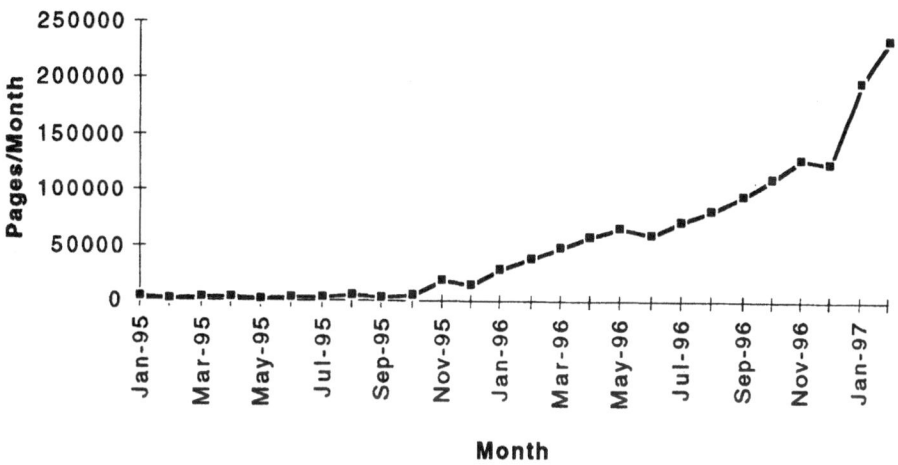

Figure 8. Retrieved article pages per month. In 1995, only the ApJ Letters were on line. Late in 1995, we started adding other journals.

older articles are read mostly in the archive and found through searches.

4. Mirrors

In order to provide better service to non-US users, we are trying to duplicate our database in different places. We currently have mirrors at the CDS in France and at the Astronomical Data Analysis Center (NAOJ) in Tokyo, Japan. This greatly improves access times from Europe and East Asia.

The data and software are mirrored over the network every time they are updated. Semi-automatic procedures make the mirroring operation fairly easy and quick.

The ADS software was developed with the possibility of mirrors in mind. All parts of the software that depend on the location of the server and on the directory structure are guided by two configuration files. These configuration files are read during startup of the server. No re-compilation of the software is necessary to move the ADS data and software to a mirror server.

5. Future Plans

In the near future we plan to complete the scanning of the astronomical literature from 1975 to present. There are still several important journals in the planetary and solar system sciences that have not given us permission to scan their journal. Hopefully we will receive permission from them to

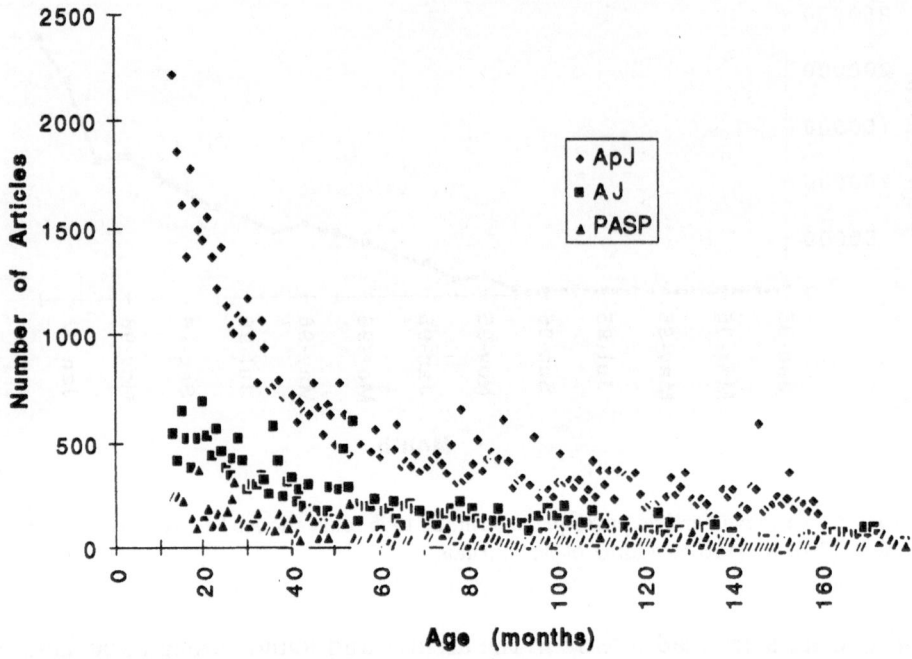

Figure 9.   This diagram shows the number of retrieved articles as a function of the age of the article. The usage of articles declines approximately with $1/t$. The plot starts at Age 12 (one year) since some journals can be made available only one year after publication. The plot for the ApJ stops at about Age 160 (about 13 years) since we have this journal only on line back to 1983.

complete the digital library in Astronomy.

In the longer term we hope to scan the historical literature as well. This is a much larger and more complicated effort and needs to be done in cooperation with the traditional astronomy libraries.

For the abstract service we plan to improve the coverage of the astronomical journals. For several journals we have only references, not abstracts. It should be in the interest of the journals to be represented in the main search system for the astronomical literature. Any journal that is not represented with abstracts will per force not be represented well in the references returned to our users. Since our users comprise the vast majority of the astronomical community, this will be a significant disadvantage for these journals.

We are currently working with the publisher of the IAU Circulars and the Minor Planet Center to include their publications in the ADS. These would be included as soon as they are published (within less than 1 hour). We are also working with the Information Bulletin on Variable Starts to index their publication in the ADS.

We hope to be able to mirror the ADS at a few more sites. Main targets for this would be Australia, India, Germany, and Great Britain.

The collaboration with the journals of the American Astronomical Society to cross-link between the journal references and the ADS abstracts is working very well. Hopefully we will be able to start similar cooperations with other journals as they prepare to come on-line.

This overall collaboration between all the parts of the astronomical community provides the basis for an example of how a complete digital library can work. Hopefully the continuation of this collaboration will provide the astronomical researchers with a constantly improving tool for their research.

6. Addresses

The ADS Homepage is at: http://adswww.harvard.edu.

The abstract service can be accessed from:
http://adsabs.harvard.edu/ads_abstracts.

This form has links to both the primary site at SAO and to the mirrors in France and Japan.

The ADS can be contacted by sending email to ads@cfa.harvard.edu

Acknowledgment

The ADS project is funded by NASA under cooperative agreement NCCW-0024. The principal team members of the ADS are Dr. Stephen S. Murray (Principal Investigator), Dr. Guenther Eichhorn (Project Scientist), Dr. Michael Kurtz (Astronomer), Dr. Alberto Accomazzi (Computer Specialist), Carolyn S. Grant (Computer Specialist), and Elizabeth Bohlen (Computer Specialist). Special thanks to Joyce Watson, who was mainly responsible for building the synonym list.

References

1. Accomazzi, A., Grant, C.S., Eichhorn, G., Kurtz, M.J. & Murray, S.S. 1995, ADS Abstract Service Enhancements, in *Astronomical Data Analysis and Software and Systems IV*, Eds. R.A. Shaw, H.E. Payne & J.J.E. Hayes, *Astron. Soc. Pac. Conf. Ser.* **77**, 36-39
2. Accomazzi, A., Grant, C.S., Eichhorn, G., Kurtz, M.J. & Murray, S.S. 1996, The Article Service Data Holdings and Access Methods, in *Astronomical Data Analysis and Software and Systems V*, Eds. G.H. Jacoby & J. Barnes, *Astron. Soc. Pac. Conf. Ser.* **101**, 558-561
3. Boyce, P.B. & Dalterio, H. 1996, Electronic Publishing of Scientific Journals, *Physics Today*, **49-1**, 42-47

4. Egret, D, Wenger, M. & Dubois, P. 1991, The SIMBAD Astronomical Database, in
    Databases and On-line Data in Astronomy , Eds. D. Egret & M. Albrecht, Kluwer,
    Dordrecht, 79-88
5. Eichhorn, G. 1994, An Overview of the Astrophysics Data System, *Experimental
    Astronomy*, **5**, 205-220
6. Eichhorn, G., Accomazzi, A., Grant, C.S., Kurtz, M.J. & Murray, S.S. 1995a, Access
    to the Astrophysics Science Information and Abstract System, *Vistas in Astron.* **39**,
    217-225
7. Eichhorn, G., Murray, S.S., Kurtz, M.J., Accomazzi, A. & Grant, C.S. 1995b, The
    New Astrophysics Data System, in *Astronomical Data Analysis and Software and
    Systems IV*, Eds. R.A. Shaw, H.E. Payne & J.J.E. Hayes, *Astron. Soc. Pac. Conf.
    Ser.* **77**, 28-31
8. Good, J.C. 1992, Overview of the Astrophysics Data System (ADS), in *Astronomical
    Data Analysis Software and Systems I*, Eds. D.W. Worral, C. Biemesderfer & J.
    Barnes, *Astron. Soc. Pac. Conf. Ser.* **25**, 35-43
9. Heck, A. 1997, Electronic yellow-page services: The Star*s Family as an example of
    diversified publishing, this volume
10. Kurtz, M.J., Karakashian, T., Stern, C.P., Eichhorn, G., Murray, S.S., Watson,
    J.M., Ossorio, P.G. & Stoner, J.L. 1993, Intelligent Text Retrieval in the NASA As-
    trophysics Data System, in *Astronomical Data Analysis and Software and Systems
    II*, Eds. R. Hanish, R. Brissenden & J. Barnes, *Astron. Soc. Pac. Conf. Ser.* **52**,
    132-136
11. Mermilliod, J.-C., Weidmann, N. & Hauck, B. 1996, The Lausanne Photometry
    Server on the World Wide Web, *Baltic Astron.* **5**, 413-416
12. Murray, S.S, Brugel, E.W., Eichhorn, G., Farris, A., Good, J.C, Kurtz, M.J., Nousek,
    J.A. & Stoner, J.L. 1992, The NASA Astrophysics Data System: A Heterogeneous
    Distributed Processing System Application, in *Astron. from Large Databases II*,
    Eds. A. Heck & F. Murtagh, *ESO Conf. & Worskops Proc.* **43**, 387-391
13. Plante, R.L., Crutcher, R.M. & Sharpe, R.K. 1996, The NCSA Astronomy Digital
    Image Library, in *Astronomical Data Analysis and Software and Systems V*, Eds.
    G.H. Jacoby & J. Barnes, *Astron. Soc. Pac. Conf. Ser.* **101**, 569-572
14. Schmitz, M., Helou, G., Dubois, P., LaGue, C., Madore, B., Corwin Jr., H.G. &
    Lesteven, S. 1995, NED and SIMBAD Conventions for Bibliographic Reference Cod-
    ing, in *Information & On-line Data in Astronomy*, Eds. D. Egret & M.A. Albrecht,
    Kluwer Academic Publ., Dordrecht, 259-270
15. Squibb, G. 1987, NASA Report on Astrophysics Data System Workshop
16. Squibb, G. 1988, NASA Astrophysics Data System Study, Final Report

ELECTRONIC YELLOW-PAGE SERVICES:

THE STAR*S FAMILY AS AN EXAMPLE OF DIVERSIFIED PUBLISHING

A. HECK
Observatoire Astronomique
11 rue de l'Université
F-67000 Strasbourg, France
heck@astro.u-strasbg.fr
http://cdsweb.u-strasbg.fr/~heck

Abstract.
    The broad currently accepted definition of electronic publishing encompasses also yellow-page services on the web. Together with their equivalent on paper, they are an example of diversified publishing. Some of these include validation and authentication steps which are *sine qua non* requirements for a service worthy of its name. We briefly describe here such a service with its outstanding features and procedures. We also discuss the maintenance processes and illustrate how constraints at the level of the distribution can downgrade an otherwise rich compilation of information.

1. Introduction

Putting a document on the WWW or setting up an information resource on the web is now considered as an act of 'electronic publishing' (see e.g. Heck 1997 and references therein). This is therefore the case for yellow-page services now accessible via the networks. When the corresponding information is also available on other media (such as paper), we can then speak of 'flexible publishing' or, as we prefer it, of 'diversified publishing'.

    In the following, we describe and comment an example of yellow-page services, the *Star*s Family*, that involves authentication and validation steps. We shall also discuss the maintenance of such a resource and indicate that the information available in the master files and on paper is actually

Astrophysics and Space Science **247**: 211–220, 1997.

richer than that provided electronically, with potentialities not (yet) fully exploited on the web.

The *StarPages* (`http://cdsweb.u-strasbg.fr/starpages.html`) cover the products of the *Star*s Family* that are available through the WWW server of the *Centre de Données astronomiques de Strasbourg (CDS)* (see e.g. Egret & Genova 1997). The *Star*s Family* itself is a growing collection of directories, dictionaries, databases and related products (Heck 1994). Nobody today would question the usefulness of telephone books nor that of remotely accessible databases. The *Star*s Family* combines their advantages by offering resources of detailed information, both on paper and on-line, about astronomical organizations, as well as on services of general and practical utility for astronomers and related scientists, and on these persons themselves.

The outstanding features include:

- a long (over twenty years now) tradition/experience in compilations of this kind;
- a resulting excellent exhaustivity of entries (including also thousands of entries without an WWW link yet[1] (directory *Starguides* and database *StarWorlds*);
- a homogeneous coverage of all practical data;
- a permanent updating and quality checking scheme carried out humanly (or 'manually') and including authentication of data originators as well as a critical independent evaluation of the information;
- the largest amount of URLs or WWW links available in a set of astronomy resources (about 7,000);
- a unique collection of (about 110,000) real acronyms and abbreviations extracted from scanned literature[2] (dictionary *StarBriefs* and database *StarBits*);
- a highly successful database of personal web pages of individual astronomers and related scientists (database *StarHeads*).

The basic philosophy of these directories and databases is to provide practical data which one seeks always to have at one's disposal. They have proved over the years to be extremely valuable auxiliaries.

---

[1]Contrary to most on-line resources, the *Star*s Family* products are not only e-mail or WWW-oriented.

[2]In other words, they are not (as in some on-line acronym servers) built up automatically from dictionaries.

2. Presentation of the resources

2.1. STARGUIDES AND STARWORLDS

The compilation of data on astronomy, space sciences and related organisations was initiated by us at the end of the seventies with the publication of directories of astronomical associations and societies, as well as of professional institutions. The two lists were merged later into a single work called *StarGuides* (Heck 1993a). From the start, the geographical coverage was world-wide and the fields covered were progressively broadened to include all organizations that could be of interest, to any degree, for astronomers and related scientists.

StarGuides' master files gather together all practical data available on associations, societies, scientific committees, agencies, companies, institutions, universities, etc., or more generally organizations, involved in astronomy and related sciences. Many other types of entries have also been included such as academies, bibliographical services, data centres, dealers, distributors, funding organizations, IAU-adhering organizations, journals, manufacturers, meteorological services, national norms and standards institutes, parent associations and societies, publishers, software producers and distributors, and so on.

Currently about 6000 entries from about 100 countries have been selected. For each entry, all practical data available are listed. Refer to the bibliography and the on-line documentation for details.

StarGuides' master files have been made accessible as an on-line database called *StarWays* by the European Space Information System (ESIS) group (see e.g. Heck *et al.* 1992) and as an independent database called *StarGates* (Albrecht & Heck 1994b) at the European Southern Observatory (ESO).

Since January 1994, the *StarGuides* files can be queried through the WWW server of the Centre de Données astronomiques de Strasbourg (CDS) as the database *StarWorlds* (Heck *et al.*, 1994).

Refer to the on-line documentation for tips on how flexible queries can be performed, often calling for subliminal (hidden) information such as synomyms (such as names of main cities in various languages, and so on) or for categories of entries corresponding to the thematic subindices of the paper versions. When retrieved, active URLs allow straight navigation towards the corresponding organizations.

2.2. STARHEADS

As a complement to the previous resource and with the development of the WWW, we started compiling a database of URLs of personal pages of individual astronomers and related scientists. That resource, called *StarHeads* (Heck 1995), has also been made operational in January 1994 on the CDS WWW server.

At the time of writing, more than 2500 personal pages are accessible, but that figure is increasing weekly as the resource is extremely successful, frequently visited and pointed at by services such as NASA's ADS (see e.g. Eichhorn 1997).

Refer to the on-line documentation for querying procedures and additional details.

2.3. STARBRIEFS AND STARBITS

The dictionary *StarBriefs* (Heck 1993b) gathers together currently about 110,000 abbreviations, acronyms, contractions and symbols. Many entries in common use and/or of general interest have also been included when appropriate.

The underlying idea is to offer to astronomers and related scientists a practical assistant in decoding the numerous abbreviations, acronyms, contractions and symbols that they might encounter in their professional activities. Maybe a bit paradoxically, if scientists can quickly grasp the meaning of an acronym purely in their specific field, they will probably have more difficulties with adjacent fields. It is actually for this purpose that this dictionary might be more often used. Scientists might also use this compilation to avoid assigning an acronym that already has too many or confusing meanings.

This compilation is essentially carried out in parallel with the permanent updating of *StarGuides/StarWorlds*. In practice, all major abbreviations and acronyms encountered when scanning the general literature and the documentation received in relation with these are gathered, the underlying principle being that they might also appear one day under the eyes of astronomers and related scientists.

The dictionary *StarBriefs* has also been made accessible as an on-line database at ESO under the label *StarWords* (Albrecht & Heck 1994a). Since January 1994, it is reachable on the CDS WWW server as the database *StarBits* (Heck *et al.* 1994).

Refer also to the on-line documentation for querying procedures and additional details.

3. Maintenance and Quality

The *Star*s Family* compilations have taken advantage of the experience gained over the years, especially in the development of techniques for collecting, verifying and treating the data. To compile a directory or a database of real value is indeed quite a different venture compared to barely reproducing and distributing, with comments of greater or lesser interest, data collected indiscriminately from all available sources. The latter criticism also applies to information gathered from on-line forms which enter the information into on-line resources without much further processing nor appropriate perusal.

Basic checks, homogenization, validation of the substance itself as well as of its (sometimes electronic) originators, and so on, are fundamental processes for a reliable product of quality. Moreover, while professional file and database construction techniques are necessary, they cannot save the extensive background, unrewarding and very careful work which is indispensable for the compilation of a valuable resource.

The definition of a very well-profiled and adapted questionnaire, the homogenization of the data collected and the maximum reduction of the respondents' biases are all points that must be satisfied, often with the help of the most modern communication means. The continuous political evolution of the world has also to be taken into account. If the information is provided in the *Star*s Family* files *bona fide*, the best effort is nonetheless made to keep track of the modifications happening and to implement them as soon as they are confirmed or recognized by the international community. On a more pragmatic note, the frequent changes occurring in phone/fax numbering in countries round the world can only be echoed in the resources through efficient collaborations with the corresponding telecommunications companies. The same applies to organizations linked to the *World Meteorological Organization* and the *International Organization for Standardization*.

One can never stress enough the importance of this obscure daily work, consisting of patiently collecting data, checking and re-checking information, and continually updating the master files. If scientists have a natural tendency to design projects and software packages involving the most advanced techniques and tools, there is in general less enthusiasm for the painstaking and meticulous long-term maintenance which builds up the real substance of the databases. This has also to be carried out by knowledgeable scientists or documentalists and cannot be delegated to inexperienced clerks.

The fashion is now shifting towards designing and testing quality control processes, but we believe that the best quality assurance (accuracy, homogeneity, exhaustivity, ...) has to be achieved when collecting and entering the data themselves with an immediate check of the entered material. None of the algorithms currently available has really convinced us of its absolute

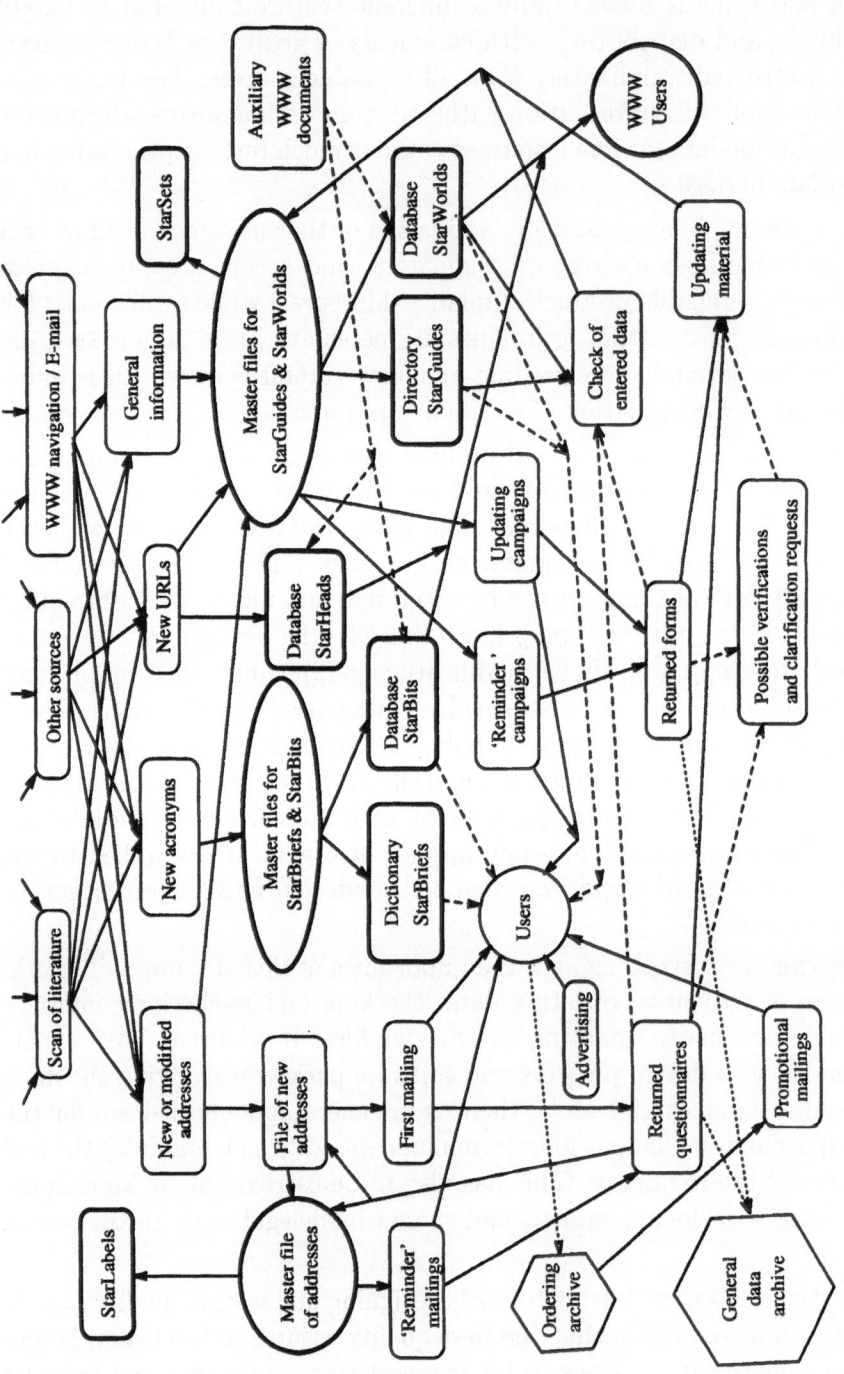

Star*s Family & StarPages General Working Scheme

(A. Heck - April 1997)

necessity and satisfactory utility. Again here, developing such processes is an appealing challenge for scientists, but most of the algorithms designed work statistically. For a database user, it does not matter much whether it is accurate up to 90% or 95%. The user wants to find the piece of information he/she is looking for, and, if found, this has to be accurate. All these considerations are obvious if a phonebook is taken as a model for yellow-page services.

It is interesting to notice how questionnaire respondents are sometimes unable to fill in properly these forms about their own organizations. Some astronomers seem to still ignore the differences between minutes and seconds of, respectively, degrees and hours (a factor fifteen between these units), which – if left uncorrected – would of course offset seriously the location of the corresponding observing places! Our experience is that it is absolutely necessary to cross-check every bit of information. We therefore request as much documentation as possible (activity reports, periodicals published, and so on) to be posted to us.

Over the twenty years or so we have been dealing with this kind of activity, we have gone through quite a few cases of ghosts associations, of non-existing groups (or groups without legal existence) within universities, of fights between leading organizations about their actual size, activities, representative status, and so on. In order to release accurate information, we have then to run independent checks, call or e-mail trusted informants and request third parties to report on the actual situation. As human aspects are always involved, diplomacy is the rule, but sometimes people have to be told bluntly to behave. In some instances, entries had to be withdrawn from the databases.

A kind of intuition (maybe this is real 'experience') has been developed over the years for detecting such cases (not very frequent fortunately, but not rare either) and the most common action is often to tune down some exaggerated figures or scope of activities, etc.

The so-called 'grey' literature has been more difficult to detect with the advent of desktop publishing packages and high-quality laser printers. The WWW and the possibility for each individual to set up impressive pages makes it even more difficult to assess exactly what is behind some of these[3]. This is why, until possible further technological developments come about, we always request an independent – and properly documented – authentication of the organization and of its representative by 'snail mail'. Of course, WWW pages that cannot be accessed, that are empty, or that contain too much unrelated material are ignored.

---

[3]Remember the famous cartoon featuring two dogs keying in e-mail messages: *On the Internet, nobody knows you are a dog!*

4. Information compiled and information distributed

On the other end of the process, it is also important to monitor the usage of resources (how in fact these are queried or tackled) and possibly to adapt the release and display of the retrieved information in a more suitable manner for the average user (with some restrictions and security measures of course[4]).

Maybe somehow surprizingly, our experience with the various on-line distributors of the information we have compiled is that this is never made totally available nor fully exploited, for either technical or human reasons[5].

Thus, in the original master files, about a hundred special characters (accentuated or others) have been encoded and are properly printed via TEX in the paper versions of the directories and dictionaries. Only a small part of these are actually displayed as such on line. All the others are transliterated in basic ASCII.

As indicated earlier, active URLs allow easy navigation towards the WWW servers of the organizations whose entries are retrieved. Depending from the technical possibilities of the visiting sites, plug-in automatic phone or fax numbering could take advantage of the standard corresponding format.

More flexible queries can be developed, possibly linked to mapping facilities very useful for planning observational campaigns, for instance. The format of the coordinates for geographic locations has been developed for making feasible the retrieval of, for instance, sites within a specific area.

Developing multilingual interfaces is also feasible as a language flag is embedded in each entry.

There are thus ample possibilities for other distributors to develop varied services from the same set of master files, including also their cross-linking. From the reverse point of view, there is also the possibility for resources such as *StarWorlds* and *StarBits* to be pointed at – similarly to how ADS is pointing to *StarHeads*.

5. A Few Last Comments

The profile of the directories and the databases, as well as the questionnaires sent to the various organizations listed have been improved and adapted over the years. The information gathered has been more and more comprehensive. The categories of the entries listed have also been gradually

---

[4]Nobody really wants his/her thousands hours of compiling work to be unfairly copied by a couple of mouse clicks.

[5]In fact, it is a specific full-time job to compile information and maintain its high-quality level, distinct from making it available on the networks.

broadened to better serve the needs of astronomers and related scientists. When compiling files such as the *Star*s Family* master ones, one cannot but be impressed by the very broad spectrum of disciplines to which astronomy and related sciences are linked, and by the very large variety of techniques applied in these fields.

The successive releases of the directories and databases give fairly accurate global pictures of the active organizations in the fields covered. Their sequence testifies to the sometimes rapid evolution of scientific interests, of data collecting and handling techniques, as well as of communications in the broad sense. A few countries have also rearranged the structure of their national facilities in the course of the past years.

Among the most striking recent changes, electronic mail and the WWW have dramatically modified the way scientists communicate and exchange information.

Of course, the political evolution in the world is directly reflected in the information provided. The USSR, the German Democratic Republic and Czechoslovakia have disappeared. New countries have lengthened *Star-Guides'* table of contents. The liberalization of political regimes, especially in Eastern Europe, has resulted in a dramatic increase of the questionnaires returned from the regions concerned. Most of the African continent remains however a dramatic gap and this should be a concern for each of us.

Technical evolutions have also been playing a significant rôle in recent years and we went through them while producing the successive versions of the directories and setting up the databases. Desktop and electronic publishing, with all the indexing facilities, have become instrumental for providing monthly updated releases of the *Star*s Family* products on paper with an outstanding quality essentially due to the TeX typesetting system.

As to the databases, not only flexible management systems allow efficient information retrieval, but the current omnipresence of WWW browsers have made their access much more popular. Specific problems are not absent though, such as the (in)stability of sites, URLs and pages. This might be due however to an unavoidable running-in phase.

Acknowledgements

Special thanks are directed to all persons who have assisted us over the years in the materialization of the *Star*s Family* products at all levels. Most of their names appear in the bibliography and in the quoted papers. A special mention is due here to Daniel Egret and François Ochsenbein for making the *StarPages* accessible on the CDS WWW server and improving the service whenever they can.

We are also very grateful to all persons and organizations who contribute to the very substance of the *Star*s Family* products by returning the questionnaires, by providing the relevant documentation, by participating in the various procedures of maintenance, validation and verification of the information, or otherwise. The *Star*s Family* products have been conceived for them and for the vast community of users. We are looking forward to satisfying their needs in continually better ways.

The implementations as databases of the *Star*s Family* products by the European Space Agency, the European Southern Observatory and Strasbourg astronomical Data Centre have been strong incentives to continue and always improve these time-consuming compilations.

References

1. Albrecht, M.A. & Heck, A. 1994a, StarWords – A database of abbreviations, acronyms and symbols in astronomy, space sciences and related fields (announcement of a database), *Astron. Astrophys. Suppl.* **103**, 471

2. Albrecht, M.A. & Heck, A. 1994b, StarGates – A database of astronomy, space sciences and related organizations of the world (announcement of a database), *Astron. Astrophys. Suppl.* **103**, 473-474

3. Egret, D. & Genova, F. 1997, The rôle of data centres in the era of electronic publishing, this volume

4. Eichhorn, G. 1997, The digital library of the Astrophysics Data System, this volume

5. Heck, A. 1993a, StarGuides – A directory of astronomy, space sciences and related organizations of the world (announcement of a catalogue), *Astron. Astrophys. Suppl.* **102**, 85-86 (see also the URL: http://cdsweb.u-strasbg.fr/~heck/sgpres.htm)

6. Heck, A. 1993b, StarBriefs – A dictionary of abbreviations, acronyms, and symbols in astronomy, space sciences, and related fields (announcement of a catalogue), *Astron. Astrophys. Suppl.* **102**, 87 (see also the URL: http://cdsweb.u-strasbg.fr/~heck/sbpres.htm)

7. Heck, A. 1994, The Star*s Family: A comprehensive set of yellow-page and on-line services, *Vistas in Astron.* **38**, 401-418 (see also the URL: http://cdsweb.u-strasbg.fr/starsfamily.html)

8. Heck, A. 1995, StarHeads (announcement of a database), *Astron. Astrophys. Suppl.* **109**, 265 (see also the URL: http://cdsweb.u-strasbg.fr/starheads.html)

9. Heck,A. 1997, Electronic publishing in its context, this volume

10. Heck, A., Ciarlo, A. & Stokke, H. 1992, StarWays – A database of astronomy, space sciences and related organizations of the world (announcement of a database), *Astron. Astrophys. Suppl.* **96**, 565-566

11. Heck, A., Egret, D. & Ochsenbein, F. 1994, StarWorlds – StarBits (announcement of two databases), *Astron. Astrophys. Suppl.* **108**, 447-448 (see also the URLs: http://cdsweb.u-strasbg.fr/starworlds.html & http://cdsweb.u-strasbg.fr/starbits.html)

ELECTRONIC PUBLISHING AT INSTITUTE OF PHYSICS

A. DIXON
Institute of Physics Publishing
Techno House
Redcliffe Way
Bristol BS1 6NX, UK
^{TM}*dixon@ioppublishing.co.uk*

1. Introduction

Institute of Physics (IoP) is one of the leading supporters of physics, both as a discipline and as a profession, on the international stage. The Institute has its origins in the Physical Society of London, formed in 1874. In 1918, in response to demands for professional representation from physicists employed during the first World War, the Institute itself was founded as a professional and qualifying body.

The aim of the Institute, as stated in the 1970 Institute of Physics Royal Charter is to "promote the advancement and dissemination of a knowledge of and education in the science of physics, pure and applied".

Institute of Physics Publishing (IoPP) has its origins in the Proceedings of the Physical Society, first published more than 120 years ago. In the mid-1960s the Proceedings were split into three to create the core of the Journal of Physics series, which remains one of the world's leading sources of physics literature. Institute of Physics Publishing is a not-for-profit publisher wholly owned by Institute of Physics and is responsible for some 5% of the world's physics literature. In 1990 and 1995 Institute of Physics Publishing was granted the Queen's Award for Export Achievement and continues to derive more than 80% of its revenue from overseas.

Today Institute of Physics Publishing employs some 170 people in Bristol, UK and Philadelphia, USA; in 1996 its turnover was £15M. The electronic publishing team comprises just 8 people, split almost equally between Programmers and Producers; between them they create all the electronic products and services offered by Institute of Physics Publishing. In 1996

Astrophysics and Space Science **247**: 221–239, 1997.
© 1997 *Kluwer Academic Publishers.*

the Institute of Physics/Institute of Physics Publishing Web Site served over 6,000 pages (not hits) each day and in 1997 it is servicing some 30,000 pages per day. It is quite possibly the most visited physics publishing site in the world. Institute of Physics Publishing handles internet connectivity for Institute of Physics and Institute of Physics Publishing, using, currently, a 512kb/s line to Telehouse in London and using Pipex as a service provider.

2. History

Electronic Publishing (EP) commenced in earnest at Institute of Physics Publishing in 1989 when we began to encourage authors to submit files electronically. In 1996, and as an average across the 31 journals published, 50% of files were received in TeX/LaTeX, 21% in word processed or ASCII files, 2% as Camera Ready Copy, and 27% as paper manuscripts. There are however significant differences in this breakdown by journal, with some enjoying TeX/LaTeX electronic file submission rates of up to 85%. This gave us a great advantage when it came to Electronic Publishing. To this day all files are converted into TeX/LaTeX, with graphics being converted into TIFF files and then 300/600 dpi (dots per inch) EPS files. Content is held within different parts of the Production system as TeX/LaTeX, PostScript and PDF files. In 1996 64,000 pages were published for Electronic Journals.

In 1993 Institute of Physics Publishing embarked, with the British Library, upon Project Elvyn[1] The objectives of the project were to establish the pitfalls and possibilities of Electronic Publishing. Six UK universities were chosen for this pilot. It became apparent that although electronic journals were desirable, for both researchers and information professionals, that the sites chosen had considerable hardware and software deficiencies and inconsistencies.

In January 1994 Institute of Physics Publishing entered into what we hoped to be the first of many joint-publisher initiatives. With Elsevier B.V. of the Netherlands we set up CoDAS, *Condensed Matter Alerting Service*. Initially it was an FTP (file transfer protocol) service run by Elsevier and only containing Institute of Physics Publishing's and Elsevier's relevant journals. This was soon to change as we shall later see.

In September 1994 the Editorial Board of Institute of Physics Publishing's *Classical and Quantum Gravity* (CQG), a forward-thinking group of technophiles, encouraged us to launch the first Electronic Journal in physics. Thus Classical and Quantum Gravity became available on Go-

---

[1]See "Project Elvyn: an experiment in electronic journal delivery. Facts, figures and finding" edited by Fytton Rowland, Cliff McKnight & Jack Meadows, published by Bowker Saur (ISBN: 1-85739-161-6).

pher, Listserv and the shiny new World Wide Web. Indeed for the first few months of its life the Web version of Classical and Quantum Gravity was the least used. Classical and Quantum Gravity electronic version was made available at no extra charge to existing subscribing institutions. Since that time Classical and Quantum Gravity has increased submissions, increased subscriptions and has even higher quality levels. Fig. 1 shows the Classical and Quantum Gravity homepage now.

Finally in 1994 we launched the *Physics World Jobs* e-mail service. As many other companies have discovered, these services are very popular and for the first 18 months of its life it was rarely out of the Top 3, in terms of Web pages served. This service has since been continually updated and improved.

1995 saw Institute of Physics Publishing crack the problem of elegant conversion from LaTeX to HTML, including maths (some 70% of Institute of Physics Publishing serials output contains complex maths). This manifested itself in *Physics Express Letters* (PEL), a service which offers full HTML, and PDF and PostScript of rapid communications from the 12 Institute of Physics Publishing journals which carry letters. It was and is a free service. In the first 3 months alone 3,000 users registered. Publication is 5-12 weeks from submission; this compares very favourably with our competitors. Fig. 2 shows a page from this service.

Also in 1995 we commenced work on electronic projects not directly related to serials. Thus we witnessed the birth of *PhysicsNet*, an online Buyer's Guide (which, incidentally, has now supplanted its original paper medium) funded by advertising and sponsorship. This provides invaluable comparative purchase information for buyers and is free to users. Fig. 3 shows a page from this service. Obtaining advertising and sponsorship revenues has proved difficult, but possible.

Finally in 1995, in terms of products or services launched, we updated *Physics World Jobs* so that it became a Web offering. Redesigned to make it easier to use, with a search engine added and the latest jobs flagged, this service increased its already high popularity levels. It can be seen in Fig. 4.

All in all in 1995 we served almost exactly 1M pages from the Institute of Physics Publishing Web site, which was completely redesigned to encompass new offerings, improve navigation and layout, and offer the beginnings of an intuitive interface.

1995 was also the year when all the research, planning and hard work occurred which finally led to Institute of Physics Publishing becoming the first major publisher in the world to offer all of its serials output online. Research commenced in the Spring with some 14,000 researchers in physics and librarians being asked about their Electronic Publishing needs and

Classical and Quantum Gravity

Search all available issues

Latest issue: Volume 14, Number 2, February 1997
(L29-L48, 257-576)

Forthcoming Articles ★ NEW

Volume 14, 1997
- Number 2, February (L29-L48, 257-576)
- Number 1A, January (A1-A344)
- Number 1, January (L1-L28, 1-256)

Volume 13, 1996
- Number 12, December (L135-L150, 3121-3288)
- Number 11A, November (A1-A316)
- Number 11, November (L125-L134, 2865-3120)
- Number 10, October (L117-L124, 2609-2864)
- Number 9, September (2329-2608)
- Number 8, August (L95-L116, 2041-2328)
- Number 7, July (L87-L93, 1691-2039)
- Number 6, June (L67-L86, 1279-1690)
- Number 5, May (L41-L66, 799-1278)
- Number 4, April (L33-L40, 575-798)
- Number 3, March (L29-L32, 321-574)
- Number 2, February (L13-L28, 161-320)
- Number 1, January (L1-L12, 1-160)

Journal Archive
- Volume 12 (1995)

This Journal's Home Page
All the latest information about this journal for authors,
referees and readers. Also available are full text featured
papers, Editorial Board details and Notes for Authors.

Figure 1.   Classical and Quantum Gravity homepage

 **PHYSICS EXPRESS LETTERS**

Welcome to the newest version of Physics Express Letters

This free service gives you immediate access to the full text and abstracts of all Letters and Rapid Communications published in 12 Institute of Physics Publishing journals. Please select one of the available access options below:

1. Physics Express Letters (*Standard Service*)

- Provides access to all tables of contents, abstracts and full text in Acrobat PDF or PostScript or HTML formats.
- No username or password required.
- Personalization Options not available.

2. Physics Express Letters with Personalization Options (*Enhanced Service*)

- Same standard features as option 1.
- Plus Personalization Options: filing cabinet, personal main menu, e-mail alerting service, personal default searches and configurable PostScript downloads.
- Username and password required: you need to create a username and password before choosing this option. Lost password assistance is available.

We would like your help

Our electronic publishing team is constantly enhancing existing products as well as developing new services. Any feedback or comments that we receive from anyone interested in physics on the Web helps us to find out what you want.

It would help us considerably if you would complete a small online questionnaire which could provide us with the information we need.

Completing the form will take no longer than 2 minutes and any information that you **Online Help** be used for market research purposes.

Figure 2.   Physics Express Letters page

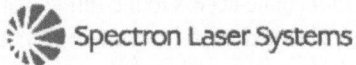

Main Menu

Buyer's Guide
Global Search: Search the Whole Database
Sponsors: Detailed information on key products ★
New Product Announcements
What's New

Other Options
About *PhysicsNet*
Information for Suppliers

Figure 3. Byuer's Guide page

wishes. From this we learnt that physicists and librarians faced many problems. For physicists, lack of time due to pressure to publish; less time to conduct research due to administrative and teaching duties; information overload, and indeed "information anxiety"; plus increased specialisation and therefore the need to monitor allied subjects. For librarians lack of funds; lack of space; a dramatic increase in published and "grey" output; uncertainty; a changing role and higher expectations from their customers. We took all this, and more, information and determined a policy.

The policy was, and is, that all Institute of Physics Publishing journals were to be available, in full, on the World Wide Web, by Spring 1996. Furthermore we determined that electronic access is included in the price of a full-rate subscription. We have since extended this concept even further

PHYSICS *world* JOBS

Main Menu

- Latest vacancies: 17/2/1997
- Search
- All current vacancies

Looking for a job in Physics? - Add PWJOBS to your bookmarks
Go straight to Physics World Jobs and search and view the latest posts, studentships and courses in physics, engineering, materials and computing.

Physics World Jobs gives immediate access - positions will appear on the site the same days as they are booked into the magazine.

So whether you are a potential employee or employer, Physics World Jobs is the service to put at the top of your bookmarks today.

Information about Physics World Jobs
- About Physics World Jobs
- Advertising in Physics World Jobs

Search

Notify Me

Figure 4.   Physics World Jobs page

in that new and existing customers in 1997 get 1996 content at no extra cost. In 1998 we shall go one step further. New and existing customers in 1998 will get full electronic files going back as far as the 1st January 1993. A five year archive for the price of a single year's subscription.

How did we come to these conclusions? There are three primary reasons:

service, support and survival.

Service

As was stated at the beginning of this chapter, Institute of Physics is a registered education charity; it would be forsaking its responsibilities if it did not play its full part in electronic communication – from the formal to the informal – helping to get information disseminated ever more widely.

Support

Institute of Physics is a non-profit making learned society, but it cannot raise all the money it needs to carry out its activities from membership dues alone. So it looks to its publishing activity to provide it with some of the resources so that it can continue to support physics and physicists.

Survival

Our publishing model is not about restricting communication, but opening it up – but we need to survive the process if we can. If it <u>can</u> be done much cheaper, or electronically, then we wish to explore these avenues.

The objectives of the Electronic Journal programme were to halt attrition (all major STM publishers face lower sales levels each year), to protect the user base, to attract authors, to increase customer satisfaction, to get to market early, to achieve a high level of usage, and not to exceed estimated costs.

In all this we were wholly supported by the Institute of Physics and Institute of Physics Publishing Boards, which are largely comprised of working physicists. This is a point which cannot be under-emphasised. Electronic Publishing is expensive and difficult and a great deal of patience is required by one's shareholders. We also ensured that customer service, sales representatives and marketing staff were fully involved in the project so that the customer and end user were always the focal point of the project.

The Project itself (now very much with a capital 'P') was run on classic project management lines and delivered on time and slightly below budget.

3. 1996

This was very much a watershed year for Electronic Publishing at Institute of Physics Publishing. Alongside Electronic Journals we also launched a further 11 electronic products and services. These catered for a wide range of end users, from authors and referees to general scientists.

Furthermore it was not just registered institutional sites which benefited from Electronic Journals. All site visitors could now see Tables of Contents of all journals, freely read the most popular papers, gather news stories, order books or reference works, read digests from Physics World or plan future conference attendance.

In April Institute of Physics Publishing took over the day-to-day running of CoDAS, now a Web service, from Elsevier. This Web service, designed and implemented by Institute of Physics Publishing, now carried 66 journals from 6 publishers and represents at least 70% of all the serials output in the world in Condensed Matter. And it costs just $95 per twelve month period for individuals. Fig. 5 shows a page from the service.

In May we created the PEERS e-mail directory which enables everyone working in science to trace their colleagues and links to e-mail and URL addresses. This carries some 11,000 names and can be seen in Fig. 6.

In July the whole Web site was again redesigned. Fig. 7 & 8 show the homepage before and after the redesign respectively. The site had grown from 600 fixed pages in 1995 to some 2,000 fixed pages, linked to a further 70,000 by cgi (common gateway interface) scripts. Along with a new corporate logo this meant that clear navigation was vital. We also took this opportunity to add several new free sections: Journal Home Pages, Schools and Colleges Service, searchable and indexed catalogues, Courses for Teachers and Technical Courses.

In September the *Author Enquiry Service* was launched. This is an electronic interface which tells an author just where his paper is in the editorial, production or publishing process. Also in this month a *Forthcoming Articles* service was made available. This provides the Tables of Contents of forthcoming articles as they become known. Both are free.

October saw the advent of online Referee Services. Now referees could complete and return reviews online, check classification codes online and keep personal information up-to-date.

During the Autumn Institute of Physics Publishing's Reference Works division also published a number of CD-ROMs including some which will, in 1997, link to Web sites which update the content (often known as 'hybrid products').

December version 2.0 of Physics Express Letters was made available, which included much of the functionality of Electronic Journals 1.2 (see below) plus extra enhancements suggested by users. Lastly, in that same month, Accelerated Publication was introduced. Now, for bi-monthly and quarterly serials publications, online readers do not need to wait for camera-ready copy. This means that content is available many weeks earlier.

Meanwhile great strides were being made with Electronic Journals.

4. Electronic Journals

There were 3 versions of Electronic Journals in 1996. The first – version 1.0 – went live on the 15th January. This offered 31 journals or 60,000 pages, accepted Class B and C IP (internet protocol) address subnets, used

Welcome to CoDAS. This service provides access to 63 journals in the fields of condensed matter and materials science. We recommend that you create a profile, as 600+ abstracts are added every week. The efficiency of your searching will be greatly improved if you work this way.

Profile
- Define Profile
- Browse or Search by Profile

Browse or Search
- All Journals
- American Institute of Physics *(Last updated: 10 February 97)*
- American Physical Society *(Last updated: 10 February 97)*
- Chapman and Hall *(Last updated: 7 February 97)*
- Elsevier Science *(Last updated: 18 February 97)*
- Institute of Physics Publishing *(Last updated: 18 February 97)*

Other Options
- Change registered details and password

CoDAS Web is a collaborative development owned by Institute of Physics Publishing and Elsevier Science.

Elsevier Science B.V.   Online Help

Figure 5.   CoDAS homepage

PEERS is a FREE service from Institute of Physics Publishing. It provides a moderated global e-mail directory of people working in science; a place where you can search for peers, colleagues or any useful contacts in your chosen scientific field.

Search the Directory

Our search engine allows you to perform a variety of searches across the PEERS directory, from simple keywords to complex Boolean expressions.

Submit Your Details

Everyone is invited and indeed encouraged to add their details to our fast-growing list. To include your details in the directory, all you need do is complete this form. A form is also available to update your details at any time.

The directory is moderated to make sure that only individuals working in appropriate positions and establishments are added to the directory.

With the service growing all the time we are keen to receive any comments or suggestions that you may have. You may want enhancements to the search engine or a facility to sort the directory by specific fields. Whatever you want, we want to hear from you - please submit your comments via our Feedback Form or contact the moderator at peers.moderator@ioppublishing.co.uk.

(PEERS is a look-up service for physicists and other scientists. It will not be used for commercial or promotional purposes, nor will any of the information submitted to the service be forwarded to other parties)

Figure 6.   The PEERS e-mail directory

▶▶▶ **Latest from The Institute of Physics**

- ☐ Visit PEERS - our new moderated, worldwide e-mail directory
- ☐ House of Lords seeks physicists' views on the 'Information Superhighway'
- ☐ Referees can now check and update their personal details online
- ☐ The Institute of Physics accelerates its online services
- ☐ Institute of Physics Publishing and Elsevier Science launch CoDAS Web
- ☐ 29 Electronic Journals now available
- ☐ Latest news and information about Electronic Journals
- ☐ Membership Information
- ☐ What's new for May
- ☐ Physics Express Letters
- ☐ *Physics World* Jobs
- ☐ Internet Resources

Figure 7.   IoP homepage before redesign

İnstitute *of* **Physics**

| New Visitors | Map | Feedback |

Online Services   Journals   Magazines   Books & Reference   Inside Physics   The Institute

Institute of Physics wins Corporate Web Site of the Year Award

Online Services

Electronic Journals
Online versions of all 31 journals published by the Institute

CoDAS Web
Over 60 journals in Condensed Matter & Materials Science

Physics Express Letters
Free online access to letters from 12 key journals ★ NEW

PhysicsNet
Free product locator service

Physics World Jobs
Free extensive job service

PEERS
Free e-mail directory

Journals

What's New
By Title
By Subject Area
Author Services
Referee Services
Subscription Information

Magazines

Physics World
Opto & Laser Europe
Scientific Computing World
FibreSystems
Astronomy & Geophysics

Books & Reference

News & Promotions
Catalogue & Ordering
Worldwide Booksellers
Author Information
Contact Details

Institute *of* Physics

About the Institute
How to Contact Us
Benefits for Members
Joining the Institute
Branches, Divisions & Groups

Institute Events

Conferences & Exhibitions
Technical Courses
Annual Congress
Courses for Teachers

Inside Physics

News in Physics
Internet Resources
International Conference Diary
Schools & Colleges

18 February 1997.
This site is best viewed with Netscape 1.1 and above.
Designed by the Electronic Publishing group.

INVESTOR IN PEOPLE   MAGELLAN 3-STAR SITE

Figure 8. IoP homepage after redesign

a simple registration form, used individual and site user IDs and passwords, provided abstracts in HTML and full text in PDF and PostScript (Classical and Quantum Gravity also still had a TEX option) and allowed for browsing and searching (including cross-journal searching).

We learnt from usage data that PDF was much more popular than our 1995 research had indicated. Indeed we now feel that formats move very quickly and that it is rather unfair to ask end users to guess what they might be using in a year's time. Users browsed twice as much as they searched (but this too is changing, see below). At an abstract level 68% went on to view the full paper (which we thought was quite high; however this figure has since remained constant). Finally didactic and pedagogical papers were, and are, popular.

Version 1.1 of Electronic Journals was published in July and here the focus was on end users. Virtual Filing Cabinets were introduced following observations of what physicists did after they had printed an article. E-mail alerting was introduced. Pre-set searching was made available. Personalization of the whole service could now be achieved. Gzipped downloads (for compression) were made easier. US paper sizes were allowed for and Featured Articles (i.e. the most popular or important in each journal) were made freely available.

After this version was published we noticed a very slight increase in PostScript use (which could have been related to the Gzip solution as its main protagonists are Unix users, who might be more likely to send a document to a PostScript printer). We also witnessed more searching (versus browsing) but this is likely to be unrelated to version improvements and merely reflecting increased sophistication by the users. Very strong growth of the service occurred between the versions, about 25% per month increases in pages served.

On 25th November Version 1.2 of Electronic Journals entered the world. Now the improvements were very much in response to the library community. Domain names were accepted (so that librarians or systems administrators didn't need to change their listing of IP addresses every time they bought or ditched a computer). All IDs and passwords were dropped (except where an individual chooses personalisation features when we of course need to know who they are) as they are, it seems, universally disliked. A remote service was introduced as we learnt that many physicists work from home or at another site on a regular basis. We also enabled users to bookmark individual journals or sections to avoid logging-in at all. A more powerful search engine was introduced. Finally (because even in November some 1997 papers were available) we enabled the 1996 archive so that 1997 customers can access it.

At the time of writing, just two months after this latest manifestation, the new features are being implemented by quite sizeable proportions of users. Remote access is the most popular, followed by the filing cabinet and e-mail alerts. Fig. 9, 10 & 11 show examples of these features.

Electronic Journals represent currently about 40% of all Institute of Physics Publishing pages served. However it is one of the fastest growing sectors. The vast majority of end users are in education/academia and are using a Netscape Navigator browser. Due to the HEFCE National Site Licence in the UK, usage is proportionately higher there; otherwise usage is very similar to print journals.

The feedback from the community has been very rewarding:

"I appreciate very much the change to strictly IP-based access to the online journals – it will make it much easier to make it available to our users, particularly students. Still offering 'enhanced access' via password is also a very good idea, and one which will probably appeal to some faculty members"
Victoria Mitchell, Science Library, Reed College, Portland, 12/11/96

"With the new version, 1.2, the access is much improved because the user is not burdened with a site id and password"
Betsy Schwartz, Brookhaven National Lab Research Library, 13/11/96

"I've gome through all recent improvements of IoP electronic journals: wonderful job! There's nothing comparable from other publishers. ...Thank you IoP very much for your efforts"
Susanna Mornati, University of Milan, 14/11/96

"I have just been having a look at the new style IoP electronic journals and am impressed. Thank you so much for taking away the need to issue yet another username and password to our users"
Lindy Wilson, Leicester University Library, 13/11/96

"Allowing access to IoP's e-journals without site id and password was an important step to go! Congratulations! This will help diminish barriers potential users might have because the access procedure is too complex."
Uta Grothkopf, European Southern Observatory Library, 14/11/96

"I want to express my appreciation of how you have enhanced and streamlined access to the electronic journals. I wish that all publishers would make it as easy for us acquisitions and serials librarians by clearly listing and distinguishing between the journals we subscribe to from those we don't! And removing the need for the individual to register for the institutional access. Thank you!!"
Karen Mokrzycki, University of California, Santa Cruz, 14/11/96

Furthermore in 1996 Institute of Physics Publishing won the Charlesworth Award for best PDF journal, was voted Most Popular Publisher by

IoP

Home | Up | Map | Feedback

Journals | Magazines | Books & Reference | Inside Physics | The Institute

Online Help

Electronic Journals

Personal Main Menu

For comprehensive, context-sensitive information on the features of this service, click on the Online Help button at the foot of any Electronic Journals page. Personalization Options are available at the foot of this page.

Search All Subscribed Journals

Selected Journals

- Journal of Physics: Condensed Matter
  Volume 9, Number 9, 3 March 1997 (L111-L144, 1889-2108)
- Journal of Physics D: Applied Physics
  Volume 30, Number 4, 21 February 1997 (499-708)
- Nanotechnology
  Volume 8, Number 1, March 1997 (1-46)
- Reports on Progress in Physics
  Volume 60, Number 2, February 1997 (151-292)
- Semiconductor Science and Technology
  Volume 12, Number 3, March 1997 (245-354)
- Superconductor Science and Technology
  Volume 10, Number 3, March 1997 (119-168)

Personalization Options

These options allow you to tailor the Electronic Journals service to match your own personal way of working. To review your current settings or to create a new set of preferences, click on the relevant option below.

- Filing Cabinet
- Personal Main Menu
- E-mail Alerting Service
- Personal Default Searches
- Configurable PostScript Downloads
- Change Your User Details and Password

Electronic Journals v1.2. Copyright © IOP Publishing Ltd. 1995, 1996, 1997. All rights reserved.

Figure 9. Electronic Journals main menu

Filing Cabinet

The **Filing Cabinet** helps you to track your favourite papers. It allows you to keep an online record of any papers that you have marked of interest so that you can quickly and easily return to them. You can also add your own personal notes that will be displayed with the paper's abstract.

To view the abstract of your filed papers, select from the list below. To add items to your cabinet, edit personal notes or remove items, use the options that appear on the abstract page.

Self-energy of a charged conducting droplet as an inversion problem
J Wehner and H J Krappe (Inverse Problems 13 No 1,)

Formulation of an adaptive sandwich beam
X D Zhang and C T Sun (Smart Mater. Struct. 5 No 6,)

Online Help

Figure 10.   Electronic Journals Filing Cabinet

the US Publishers' Communication Group, and is shortlisted for Corporate Web Site of the Year by the 1996 Corporate Publishing Awards (organised by Popular Communication Courses Ltd).

5. The Future – Short-Term

What lies ahead? In the immediate short term we have two specific goals – HyperCite^{TM} and an electronic Archive. The concept of HyperCite^{TM} is simple. Click on a reference in any paper, anywhere in the world, and read the full text of that cited reference. Fig. 12 shows a diagram illustrating the concept. The first stage of HyperCite^{TM} for Institute of Physics Publishing is to link from Institute of Physics Publishing references to the INSPEC database and from Institute of Physics Publishing papers to Institute of Physics Publishing papers. We should be ready to do this by Spring 1997. The second phase is a collaborative one with other major physics publishers so that we all link to each other. The final phase is to embrace all physics serials output.

The electronic Archive will be live on 1st January 1998. We know that

I₀P   Electronic Journals | Home | Up | Map | Feedback |
Online Services   Journals   Magazines   Books & Reference   Inside Physics   The Institute

Inverse Problems **13** No 1 (February 1997) 185-201
PII: S0266-5611(97)75932-X

Self-energy of a charged conducting droplet as an inversion problem

J Wehner and H J Krappe

Hahn-Meitner-Institut Berlin GmbH, 14109 Berlin, Germany

Received 27 June 1996

Abstract:

*The determination of fission barrier heights of charged clusters of alkali atoms
requires the calculation of the electrostatic self-energy of arbitrarily shaped,
conducting droplets. For axially symmetric shapes we propose to determine a charge
distribution on the axis giving rise to the same electric field outside the droplet as the
actual surface charge. This leads to an ill-posed inversion problem. We discuss its
regularization and demonstrate the usefulness of the method in some examples.*

Personal Notes:

Filed 3rd Feb 1997

Full Article Options:

Acrobat PDF (263k)

Gzipped PostScript (306k)

Amend Personal Notes for this Article

Delete this Article from your Filing Cabinet

Setup information for Adobe Acro Online Help pressed PostScript

---

| Main Menu | Table of Contents |

Figure 11.   Electronic Journals retrieved page

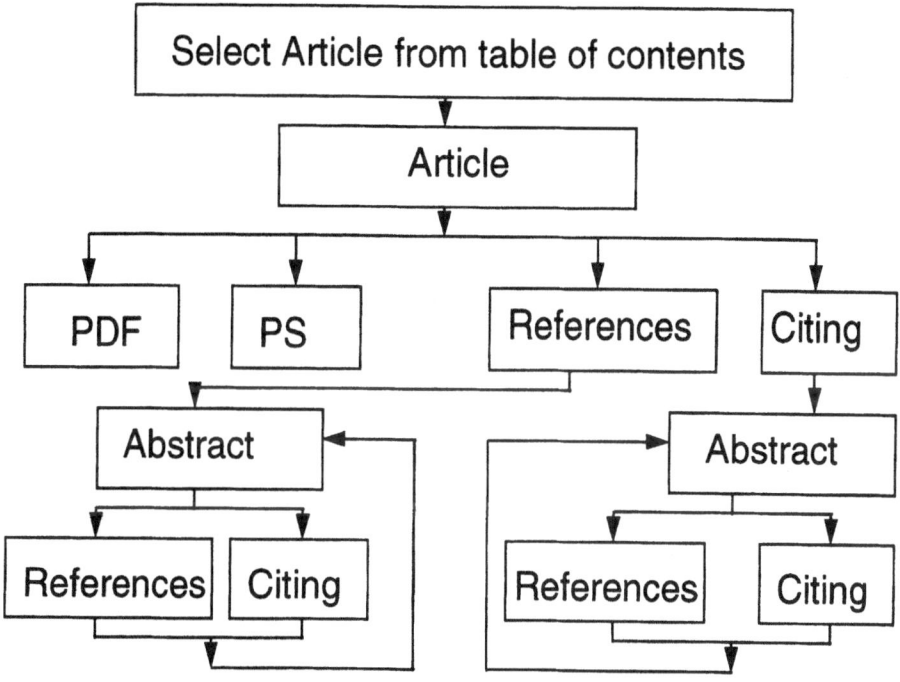

Figure 12. HyperCite^{TM} Model

this will be in high demand, following research conducted in early 1997.

6. The Future – Medium-Term

Alliances and collaborations are the only sensible way forward. As long-term maintenance costs grow in Electronic Publishing, the need to share knowledge and cut costs whilst still remaining competitive, will be paramount. Attempted communal gateways will flourish, with a few winners emerging in each segment. Interfaces will improve dramatically. New pricing models will emerge. "New" publishers will emerge. Extensions of the HyperCite^{TM} concept will flourish.

7. The Future – Long-Term

I cannot predict this. I do know that Institute of Physics Publishing will be an active and hopefully leading player.

IDEAL AND APPEAL

A Model for Consortium Licensing of Electronic Journal Collections

J. MENZEL, K. METZNER AND E. POPE
Academic Press Inc.
525 B Street, Suite 900
San Diego CA 92101-4495, USA
jmenzel@acad.com

Abstract.

In a model project unprecedented in size and scope, Academic Press has made available its 175 scientific journals to licensed institutions in full-text over the WWW since January 1996. This article describes the unique business model behind this project, and gives a first evaluation after one year.

1. Introduction

Academic Press (AP) is one of the largest commercial publishers in the United States for scientific information, and is known throughout the international scientific community. Established as a publishing company in the United States in 1942, today, Academic Press maintains locations in San Diego, London, Boston, New York, Sydney, Tokyo, and Toronto. The company publishes about 400 new books each year, and maintains a backlist of over 6000 titles in its catalog. Academic Press also publishes journals in over 30 different disciplines.

Since the advent of the World Wide Web (WWW) in 1994, many scientists have discovered the enormous possibilities this new medium provides for communicating and finding scientific information on-line in a much more timely and comprehensive manner than traditional print on paper can provide. Many publishers, including Academic Press, recognized early on the need to make their information available via the web. Combined with new procedures like electronic submission, disseminating content over the Internet can be much faster and broader than via a print medium.

Astrophysics and Space Science **247**: 241–250, 1997.

TABLE 1. Journals in physics and astronomy published by Academic Press

Journal title	1997 issues/yr
Annals of Physics	18
Atomic Data and Nuclear Data Tables	6
Icarus	12
Journal of Computational Physics	18
Journal of Fluids and Structures	8
Journal of Magnetic Resonance	12
Journal of Molecular Spectroscopy	12
Journal of Solid State Chemistry	16
Journal of Sound and Vibration	50
Journal of X-Ray Science and Technology	4
Mechanical Systems and Signal Processing	6
Nuclear Data Sheets	12
Optical Fiber Technology	4
Superlattices and Microstructures	9

Contrary to what many people believe, however, the new medium will not decrease the costs of production significantly. Only about 20% of the cost of a journal can be attributed to printing and distribution. Traditional publisher tasks, such as selection, organization, and quality control of the content will remain the same. On-line database costs only add to all that.

2. IDEAL

The *International Electronic Digital Access Library* is now in 1997 in its second year as a 3-year developmental project to bring research journals directly to the end usersÆ digital desktops. IDEAL uses open standards: End users only need an Internet connection, a World Wide Web browser, and AdobeÆs free Acrobat Reader. The full text of journal articles is displayed in the Acrobat format, tables of contents and abstracts in HTML. The system uses the Verity search engine for searching in and across journals. Academic Press currently maintains two mirrored servers, in San Jose CA and Bath UK, operated by affiliates of Fujitsu, a major international technology provider. Additional mirror sites are planned to insure efficient access from all points on the globe essentially 24 hours per day, 365 days in the year.

The online database contains information from headers supplied by typesetters in SGML format: Title, authors, abstract, bibliographic information, a unique manuscript identification number (which is also in the printed version and is part of the Publisher Item Identification or PII). APÆs typesetters use a Document Type Definition (DTD) developed by AP from available journal article DTDs. The typesetters also produce the pdf files, using the same PostScript data used for the print version. The only change is that the figures are downsampled, as a compromise between file size and quality: line drawings to 150 dpi, photographs to 300 dpi.

Since IDEAL first went "live" in January 1996, some 25,000 scientific articles published in 1996 have been posted and content continues be added at a rate of around 2000 articles per month. In this development phase there certainly are still problems galore to overcome and performance is sometimes less than optimal. It is a steep learning curve for all involved on the supplier side, publisher, typesetters, Internet hosting service provider. But there is the exhilaration of accomplishing something innovative and useful. And feedback from end users has been overwhelmingly positive. IDEAL is available at `http://www.idealibrary.com` or `http://www.europe.idealibrary.com`.

3. APPEAL

The *Academic Press Print and Electronic Access License* is a "site" license aimed at large consortia of libraries. It provides access at all sites within a licensed consortium to all the journals formerly held in print form anywhere within the consortium – all 175 Academic Press journals, if the consortium is large enough. In 1995 Academic Press signed an APPEAL agreement with a consortium of 160 British colleges and universities. In 1996 many more agreements are being signed with consortia in North America (see Table 2 for a current list), Europe, and the Far East. A consortium must have centralized purchasing authority, exercise control over passwords and IP addresses, provide some minimum level of support to the sites within the consortium, and be a willing partner with Academic Press over a three-year development period.

Anyone with Internet access can freely browse and search the journal tables of contents on IDEAL, and, for now, abstracts too. Authorized Users at sites within the licensed consortia of the project, however, can view, search, print, and download complete articles in the Acrobat format without restriction for personal use or course packs, or for internal company business purposes. APPEAL permits copying and transmitting of articles from IDEAL within the consortium, but prohibits all copying and transmitting of the electronic files outside the consortium.

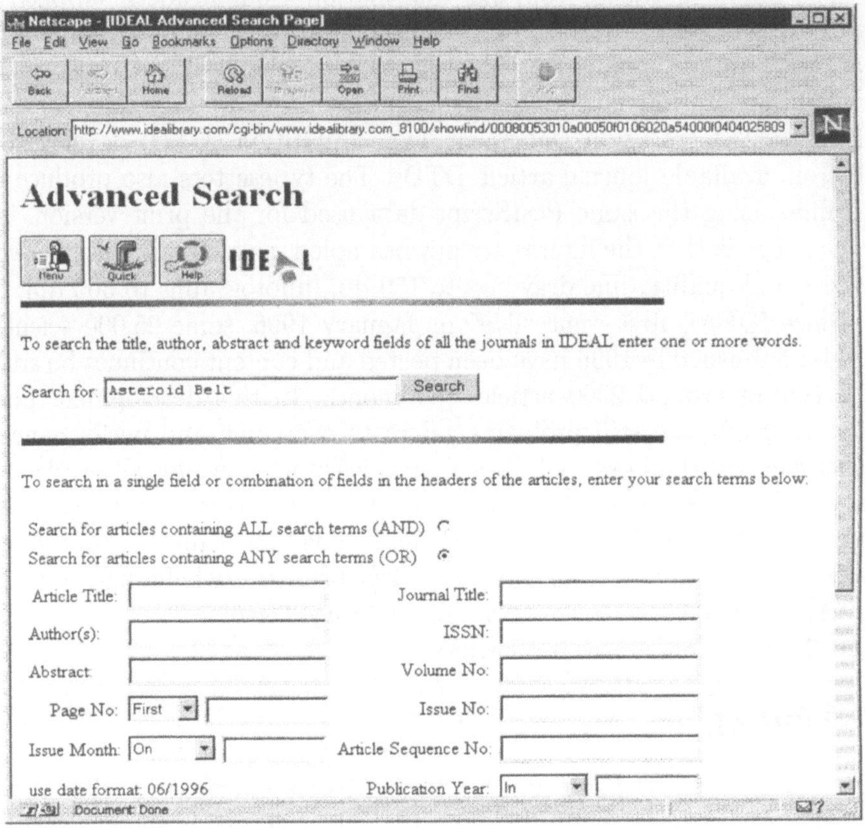

Figure 1.   TOC of Icarus (screenshot)

The license fee to a consortium for electronic access to IDEAL is based on the recent historical levels of print subscriptions within the consortium and other factors. In addition to electronic access APPEAL gives all members of a consortium the right (but not the obligation) to order print subscriptions, directly or through their customary agents, from a special deeply discounted price schedule, which averages less than 25% of the full institutional print subscription prices. The cost for print and electronic access typically will be in the range of 110-140% of print alone.

A summary version of the license agreement is included here in the Appendix. It has attained its present form as the result of extensive and intensive discussions and negotiations with interested library consortia.

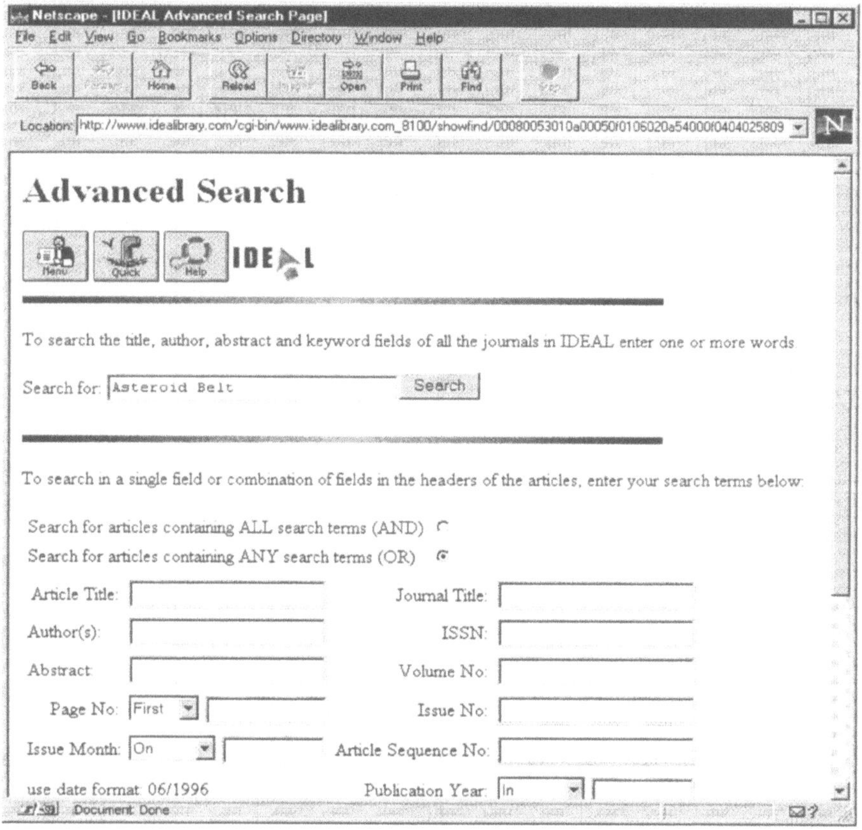

Figure 2. Advanced Search (screenshot)

4. Future

Many technical improvements and new features are under discussion, such as better worldwide and 24x7 access, linking into and out of IDEAL to other databases, better searching, supplementary material including multimedia, interest profiling, etc. These will be implemented in response to market demand and as resources permit.

A most important problem will be to develop pricing models sustainable into the 21st century based on what the journal publishing process is becoming. If there is a paradigm shift going on, it has not yet stopped shifting. The roles of all of us in the valuchain are evolving: authors, funding agencies, editors, publishers, service providers, agents and other intermediaries, libraries, and end-users. Is a journal more than the sum of its parts, more than a mere collection of articles? Is a collection or library of journals

TABLE 2. Consortia Licensed by Academic Press as of February 1997

Consortium	Number of institutions	Number of sites	Est. number of authorized users
California State University System	11	11	143,500
CIRLA (Chesapeake Information Research Library Alliance)	2	2	29,000
CISTI (Canada Institute for Scientific and Technical Information)	9	27	N.A.
GEORGIA (University System of Georgia & University Center in Georgia)	2	60	267,000
HEFCE (Higher Education Funding Council of United Kingdom)	160	160	1,740,000
NERL (North East Research Libraries)	8	11	98,900
NMLA (New Mexico Library Alliance)	3	3	N.A.
OARL (Ontario Academic Research Libraries)	7	7	148,000
OhioLINK	47	47	473,000
PICA (9 Dutch and German Universities)	21	21'	N.A.
TCLC (Tri-State College Library Cooperation)	41	41	78,000
VIVA (Virtual Library of Virginia)	15	15	354,500
Totals	326	405	> 4,000,000

a "super-journal" in some sense? What about the process by which a small niche journal becomes important and established while remaining small or becomes large and important? What about the "archive of science"?

Acknowledgements

Many people at Academic Press have been involved in the development of IDEAL and APPEAL aside from the authors of this paper, too many to list individually.

Appendix: Academic Press Print and Electronic Access License (APPEAL) – Summary, January 1997

APPEAL is a three-year licensing agreement between a group of institutions (a consortium) and Academic Press (AP), the publisher. The agreement gives researchers, faculty, students, and staff at those institutions access over the World Wide Web to the publisher/Es journals or serials. It covers the rights granted by the publisher, including the right to deep discounts on print subscriptions, the mutual obligations of the two parties, and the fees to be paid. The current full version of the agreement is the result of many discussions between AP and various libraries and consortia starting in 1995. No further changes are contemplated at this time.

A consortium is an organization that represents and consists of a group of member institutions and their libraries. Schedule IA attached to an AP-PEAL agreement lists the member institutions of the consortium and all their libraries, including all subsidiary libraries in departments, laboratories, schools, etc. A consortium should be large enough that the members collectively are subscribing to all AP journals. Schedule IB lists all the recent print subscriptions held by the members, including duplicates. Schedule II gives the calculations of the license fee (A) and the Multiplicity Factor (B). Schedule III is the complete list of publications on IDEAL.

The publisher will, of course, license industrial companies as well as academic libraries. In that case, the concept of a æconsortiumÆ may be adapted to represent a single company, possibly with multiple sites worldwide, and the list of journals in Schedule IB will usually include only journals in the areas of interest to the particular company.

The online platform or service through which Academic Press journals are made available is the International Digital Electronic Access Library (IDEAL), designed by AP and operated for AP by Fujitsu. As of December 1996 IDEAL has two mirror servers, in San Jose CA and Bath UK, and offers 173 high quality journals in full text complete from the beginning of 1996 (plus a few 1995 issues). Additional local mirror servers will be installed in other geographic areas as needed, and this may increase costs in those areas.

RIGHTS GRANTED BY THE PUBLISHER

Authorized Users may access IDEAL and retrieve, display, search, download, print, or store individual articles from the journals listed in Schedule IB for scholarly, research, educational and personal use. Authorized Users are the employees, faculty (permanent and visiting), and students of the consortium members. They do not include alumni or other organizations, companies, or institutions not explicitly included in Schedule IA. Public access may be provided to IDEAL from workstations on library premises for scholarly, research, educational, and personal use. Remote electronic access to IDEAL by members of the public is not permitted.

Copying and storing is limited to single copies of a reasonable number of individual items. Downloading an entire issue of a journal is not permitted. However, digital or print copies may be included in coursepacks and reserves, or in internal corporate training programs and drug application materials. Authorized Users may transmit downloaded copies of individual items to persons who are not Authorized Users for the purpose of scholarly communication, so long as such transmission is not done on a systematic basis. However, IDEAL may not be used, directly or indirectly, for any paid service, such as document delivery, or any systematic supply or distribution of material to non-Authorized Users, including interlibrary loans. Copyright and other notices or disclaimers by the publisher may not be removed from IDEAL or modified or obscured.

MUTUAL RESPONSIBILITIES

The publisher makes the journals listed in Schedule IB available online via the Internet at one or more IDEAL servers starting with the applicable subscription year, with the full text in Adobe Acrobat Portable Document Format (PDF) and article headers and abstracts in HTML format for viewing by Web browsers, reserving the right to change formats with six months notice; makes reasonable efforts to insure online availability of issues prior to delivery of the printed issues and to provide continuous availability, subject to periodic maintenance and updating or factors outside the publisherÆs control; assures that IDEAL performs at least as well as other similar online systems; and controls access to IDEAL by automatic IP address recognition or by passwords and usernames or both.

The consortium and its members establish and maintain Internet connectivity to the publisherÆs servers and pay all associated connection costs; obtain licenses to use the Adobe Acrobat Reader and suitable Web browsers and pay for them, if necessary; cooperate with the publisher on appropriate ways to inform Authorized Users and library patrons of the availability of IDEAL and the terms of usage; provide their valid IP addresses to the publisher; issue and terminate passwords and usernames, if appropriate; and verify the status of Authorized Users.

LICENSE FEE FOR ELECTRONIC ACCESS

The License Fee for electronic access is 110% of the Base Price. The Base Price is what the consortium would pay if the members continued their recent print subscriptions plus any new serials to be included in the license. A minimum $1000 contribution to the Base Price is required by each member or site. There is a 20% discount off the License Fee for academic consortia. The local server premium is 20%, if applicable. In subsequent years the publisher may adjust the Base Price to account for growth and other cost increases, but will limit any increase in the Base Price to at most 10% plus inflation.

New consortium members may be added at any time, starting service with any complete calendar year of the online subscriptions, provided a new member makes an additional contribution to the Base Price calculated as above from its prior print holdings, with a minimum Base Price contribution of $1,000. Note that if a new memberÆs list of journals includes titles not previously included in Schedule IB, then all consortium members now get access to these journals. On the other hand, new members with only a few titles, or even none (such as small colleges, community colleges, hospitals, local public libraries) may join and get immediate access to all the journals in Schedule IB.

Additional publications, not in Schedule IB, may be added at any time, starting service with any complete calendar year subscription. Taking into account that the whole consortium will have access to the new journal, the increase in the Base Price is the institutional print subscription price of the new journal times the consortium Multiplicity Factor (MF). The MF is the dollar-weighted average number of duplicate copies of all journals in the consortium. It is calculated by dividing the Base Price by the sum of the institutional print subscription prices of all journal titles in Schedule IB (without regard to duplicates).

Publications withdrawn by the Publisher will reduce the License Fee by the amount they contributed to the current annual fee, prorated according to the date of the withdrawal relative to the calendar year.

PRINT SUBSCRIPTIONS

The consortium, its members, and its Authorized Users may buy print subscriptions of the journals listed in Schedule IB using the Academic Press Deep Discount Price schedule (DDP) given in Schedule III. There is no obligation to subscribe: As many or as few print subscriptions as desired may be bought, directly or through the membersÆ customary agents. Note that these DDP prices vary from journal to journal, but the average for the entire list is less that 25% of the full institutional rate, a discount of over 75%.

These DDP print subscriptions may be used like any print subscriptions, consistent with applicable copyright laws, including use by non-Authorized Users and for interlibrary loans, with one exception. The exception is that they may not be resold or transferred to another institution outside the consortium.

SECURITY, EVALUATION, TERMINATION

Security: The consortium and the publisher agree to cooperate in the implementation of security protocols and procedures as they are developed during the term of the agreement. The consortium and its members agree to notify the publisher promptly of any infringements of copyrights, unauthorized use, or other misuses of IDEAL, of which they become aware and will cooperate with the publisher in investigating them. In the event of infringement by an Authorized User, the consortium or one of its members will take steps to stop it and prevent it happening again.

Analysis and Evaluation. The consortium, its members, and the publisher agree to cooperate in the collection and sharing of information about the use of IDEAL, consistent with privacy and confidentiality. They also agree to collaborate on user surveys and questionnaires.

The Term of the Agreement is three years, until 31 December 1999. It will renew automatically for additional successive one year terms unless either party notifies the other in writing not less than nine months prior to expiration. If the agreement is terminated, continued access to the material that was licensed will be provided, if requested by the consortium, in archival digital form or by ongoing online access for a reasonable fee and under the same conditions.

For further information contact one of the following:

Taissa Kusma, Director of Electronic Product Development, Chestnut Hill MA, tkusma@acad.com;

Vince Cassidy, Director of Electronic Product Development, London UK, vcassidy@apuk.co.uk.